The Question of Political Community

FRONTIERS OF THE POLITICAL

Series Editor:
Engin Isin is Professor of International Politics, Queen Mary University of London (QMUL) and University of London Institute in Paris (ULIP). He is a leading scholar of citizenship studies and is a Chief Editor of the journal Citizenship Studies. He is author and editor of eleven books in the field, including 'Being Political' and 'Citizens Without Frontiers'.

This series aims to contribute to our understanding of transversal political struggles beyond and across the borders of the nation-state, and its institutions and mechanisms, which have become influential and effective means of both contentious politics and political subjectivity. The series features titles that eschew and even disavow interpreting these transversal political struggles with categories and concepts.

Postcolonial Transitions in Europe: Contexts, Practices and Politics
Edited by Sandra Ponzanesi and Gianmaria Colpani

Citizenship and Place: Case Studies on the Borders of Citizenship
Edited by Cherstin M. Lyon and Allison F Goebel

The Question of Political Community: Sameness, Logos, *Space*
Jonna Pettersson

Postcolonial Intellectuals in Europe: Academics, Artists, Activists and their Publics
Edited by Sandra Ponzanesi and Adriano José Habed

Citizen Journalism as Conceptual Practice: Postcolonial Archives and Embodied Political Acts of New Media
Bolette B. Blaagaard

Governing Affective Citizenship: Denaturalization, Belonging, and Repression
Marie Beauchamps

The Question of Political Community

Sameness, *Logos*, Space

Jonna Pettersson

ROWMAN &
LITTLEFIELD
INTERNATIONAL
New York • London

Published by Rowman & Littlefield International Ltd.
6 Tinworth Street, London SE11 5AL, United Kingdom
www.rowmaninternational.com

Rowman & Littlefield International Ltd. is an affiliate of Rowman & Littlefield
4501 Forbes Boulevard, Suite 200, Lanham, Maryland 20706, USA
With additional offices in Boulder, New York, Toronto (Canada), and Plymouth (UK)
www.rowman.com

British Library Cataloguing in Publication Data
A catalogue record for this book is available from the British Library

ISBN: HB 978-1-7834-8891-9
PB 978-1-7834-8892-6

Library of Congress Cataloging-in-Publication Data
ISBN 978-1-78348-891-9 (cloth: alk. paper)
ISBN 978-1-78348-892-6 (pbk: alk. paper)
ISBN 978-1-78348-893-3 (electronic)

™ The paper used in this publication meets the minimum requirements of American National Standard for Information Sciences—Permanence of Paper for Printed Library Materials, ANSI/NISO Z39.48-1992.

Printed in the United States of America

Contents

Acknowledgements

There are a number of persons who have been central to the completion of this book. I especially want to mention Jens Bartelson and Sofia Näsström for their insightful comments and difficult questions. A number of persons have read and commented on various versions and parts of the manuscript. I especially owe my gratitude to Lars Toender, Gregor Noll, Johan Tralau, Carl-Göran Heidegren, Ted Svensson, Hans Ruin, Catarina Kinnvall, Kimberly Hutchings, Nisha Shah, Petter Narby, Rickard Andersson, Cecilia von Scheele, Mia Orange, Daniel Steuer, Björn Badersten, Sara Kalm, Martin Hall, and Anna Sundell. I also want to thank Ylva Stubbergaard for her generous support.

Parts of this book were written at Northwestern University, and I am grateful to James Farr for finding me a room at Scott Hall. Thanks also to Crafoordska Stiftelsen, Fahlbeckska Stiftelsen, and Lars Hiertas Minnesfond for financial support.

I am highly indebted to the helpful editors at Rowman & Littlefield, and I particularly wish to thank Engin Isin for encouraging me to proceed with this project. My sincere thank you also to the external reviewers for their very perceptive and productive comments on the manuscript.

This book has in various forms been part of my life for a long time, and I want to thank my husband, Behrang, for his support and for bearing with me during these years. Finally, I owe my gratitude to my daughter, Anoushe, whose continuous interruptions of my work have given me the distance and distraction necessary to yet again return to the text. This book is for the two of you.

Introduction

From discussions on democracy to attempts to widen the scope for citizenship beyond the confines of the nation-state, Western thinking of the political community has continued to assume the community as requiring some form of unity or unifying principle that serves as delineation.

With the understanding of political communities as bearers of particular collective identities, they are seen as the same in relation to themselves. Simultaneously they are differentiated both against each other and against an amorphous outside made up of those who are denied access to the community. This book argues that this problem stems from how Western political thinking has assumed the political community to require unity or a unifying principle as its condition of possibility. It suggests that such a unifying principle takes the form of a supposed sameness, reflected in the history, space, language, and ultimately the *logos* of the community. This sameness grounds membership in the political community and ascertains its unifying delineations. Since the sameness of one community is either established in relation to, or productive of, something different to that community, the assumption of a delineating unity at the heart of the political community inescapably binds sameness to difference. Hence, the book argues that when unity forms the condition of possibility for thinking the political community, the problem of sameness doubling up with difference will persist, as will the closures of the meaning of politics and political existence.

There is a twofold aim to this book. On the one hand, it assumes a critical perspective by showing how relying on a notion of sameness that presupposes difference as the condition of possibility of the political community will also lead to attempts to widen the scope for inclusion to give rise to new exclusions. On the other hand, it takes on a normative and hermeneutical approach through developing a notion of sameness that does not presuppose difference and a notion of difference that does not rely on an underlying assumption of identity. The book suggests that confronting political problems of exclusion and closure means that the question of political community must be both reformulated and relocated. It must be reformulated so as to grasp the meaning of an incalculable community, and it must be relocated so as to make room for a place that disrupts the persistence of spatiotemporal continuity.

The primary ambition of the book is not to pursue a conceptual analysis and approach the meaning of community through conceptual history. It rather seeks to attend to how the *thinking* of community is predicated on assumptions of sameness that presuppose difference. The kind of community stemming from such assumptions is only of secondary importance to the study. Instead, the task undertaken is to see how a notion of sameness that presupposes difference forms a condition of possibility for thinking the community in various examples of Western political thought.

So what is meant by sameness? Since the community forms an ambiguous concept with a meaning likely to alter over time and in various bodies of thought, it seems not far-fetched to think that the meaning of sameness does the same. Certainly, the meaning of sameness will comprise different aspects in different ways of grasping the political community. Shared over these contexts is, however, that sameness is formed in relation to difference and identity. Whether the community finds a unifying principle in time, space, or language, these all function to foster an identity taken to serve as an emotional pull that holds the community together. Such a unifying identity is assumed to ground the possibility that we at all can coexist in a political community and make decisions regarding its direction, ideals, and future. Now, identity is a concept harbouring an oppositional structure. It assumes the similarity between entities (i.e., their sameness), but in order to do so, it must also assume the possibility of their difference. Thus, sameness forms a concept of varying thickness. In its most extreme form, sameness is found in the ethnocultural sameness of nationalism. In its most generic form, sameness is found in the individual subject of the modern state. More than the degree to which some entities are similar, what is of interest here is the understanding of the political community and how it relates to a notion of sameness predicated on an assumption of difference.

Scrutinising different expressions of sameness reveals sameness as a problem manifested in the temporality, spatiality, and rationality of Western philosophical thinking in general and in the thinking of the community in particular. Temporal expressions of sameness assume time as a linear unfolding through which teleological progression inscribes history on a narrow path leading relatively uninterrupted from the past towards the future. Spatially, a community predicated on sameness is manifested in the homogeneous and continuous spread of land, often imbued with the linear history described by temporal sameness. While siding with the temporal perspectives of critical engagements with nationalism and exclusion (see for example Stephens 2013), this book primarily engages the spatial aspects of sameness. Thus, it seeks to discern notions of space that might disrupt the smooth space a community predicated on unity and sameness rests upon.

So why is sameness a problem? In this book, sameness emerges as a problem of closure, of boundaries, of membership, and of the possibilities of politics. Sameness and unity as a presupposition of the community creates a divide between different political communities. More importantly, it creates a divide between all beings recognised as political subjects and all those whose position with respect to the political community makes them go unregistered on the political radar. The latter—which could for example be the stateless, the undocumented migrant, the refugee who challenges death crossing the Mediterranean border to Europe, or the Roma beggar in the streets of Stockholm who has the freedom of movement according to EU law but lacks recognition as a good citizen-subject—are all left unaccounted for by the political community. Those whose lives are most exposed to or shattered by the dark sides of the nation-state order are also left out from being perceived as political beings in themselves. Their predicament might be political, but they lack appreciation as political actors and are silenced in the political processes determining their lives.

Providing the grounds for unity, the sameness of collective identities exercises a closure of both the meaning and reach of the community. Collective identities function "as ways of patterning and arranging the world that allow some people and groups to enjoy a semblance of sovereign agency at others' expense" (Markell 2003, 5). Inherent to sameness there is a closure that establishes the community as bounded and determines in advance the parameters to which one needs to conform. A notion of sameness predicated on difference is productive of boundaries that govern a physical inside from a physical outside, while also distinguishing externalities internal to the community. Boundaries of sameness separate an 'us' from a 'them', distinguish the similar from the different, and determine the practical reach and normative meaning of the political realm.

A community predicated on sameness rests on a number of closures that emerge in all aspects of the community where the assumption of sameness becomes manifest. They are seen in its borders, which, as Étienne Balibar points out, are "situated everywhere and nowhere" (Balibar 2002, 78). The boundaries and limits of community are not merely at the border but vacillate and are polysemic so that borders neither function equally to the people exposed to them, nor are they found only at the edges of territory. Borders are also inner and invisible boundaries, meaning that for those lacking formal inclusion in their state of residence, a visit to the doctor or dentist, a train ride, or a movie rental might involve an encounter with a border.[1] Hence, borders constitute less of a distinction between an inside and an outside of the state than they are constitutive of both external spaces internal to the state and spaces of state power external to the state.

A community predicated on sameness forms also a closure in terms of turning communal belonging into a question of membership. This means that a number of universal democratic and humanist values in effect become conditioned on whether one upholds such membership. With the question of membership, the problem of a community predicated on a notion of sameness presupposing difference begins to rub off also on those formally included in the community, since it suggests there being a number of features that one needs to conform to in order to be perceived as a member. Therefore, a community predicated on sameness forms a closure of what it means to be political as well as of the very meaning of politics, as it asserts that there are certain acceptable ways by which one can engage politically.

Often, attempts to relax the above problems eventually fail escaping them, since they assume unity as a condition of possibility for the community. Through elaborating what is here called "incalculable space" and through attending to the meaning of place, this book seeks to grasp an understanding of a community that escapes being predicated on sameness, difference, and identity; that is, a community *where there is place for all.*

Before proceeding to do so, light should be shed on the theoretical and practical import of the problem of sameness.

I

To grasp how sameness operates in relation to identity and difference, Paul Ricoeur's distinction between sameness as *idem* and selfhood as *ipse* as two aspects of identity can be helpful. Sameness, *idem*, is a concept of relation and connotes numerical and qualitative identity. Numerical identity refers to unicity and finds its opposition in plurality. The expression "one and the same" would apply to numerical identity, since it does not understand two things bearing the same name as two different things but rather as identical. This means that numerical identity turns cognition into recognition and identification into reidentification through allowing the same thing to be counted twice. By contrast, qualitative identity refers to resemblance and is contrasted to difference. It describes an extreme resemblance so that two components are "interchangeable without any noticeable difference" (Ricoeur 1994, 116).

Whereas sameness, or *idem*-identity, is relational in between entities or beings, selfhood, or *ipse*-identity, is grasped with respect to the individual. Selfhood describes the capacity of agents to think of themselves as authors of their actions and bearers of beliefs. *Ipse*-identity is reflexive to its constitution and describes the possibility of an agent to reflect on herself and on her actions. *Idem* and *ipse* are not at all moments kept apart. Especially in the concept of *character*, they tend to overlap or be conflated with one another. To Ricoeur, character describes a "set of distinctive marks which permits the reidentification of a human individual as being the same" (ibid. 119) and "designates a set of lasting dispositions by which a person is recognized" (ibid. 121). In the notion of character, selfhood emerges as sameness; even if the character I talk of refers to me myself, it announces itself as *idem*. Hence, rather than asking the question of "who?" the notion of character will turn identity into a question of "what?" (ibid. 121–22).

Although Ricoeur largely discusses personal identity, Hans Lindahl suggests *idem*- and *ipse*-identity to be applicable also to collectives. *Idem*-identity provides for a closure of the collective through establishing it as the same in relation to itself while different with respect to other collectives. To Lindahl, the closure performed by *idem*-identity leaves room for alterity and plurality and forms therefore not the primary ground for inclusion and exclusion. Since inclusion and exclusion are actor-categories, he argues, they are made sense of through selfhood, or *ipse*-identity. Applied to a group, selfhood draws the *collective* as an agent. Accordingly, the collective identity emerges not only as sameness, but also as *ipseity*, as a collective self that turns inclusion into *self*-inclusion (Lindahl 2014, 91–92).

While Lindahl's discussion concerns legal collectives, it is instructive for the sameness at work in the thinking of community. It points at how the self of the first-person singular perspective, of 'I', is transposed to the first-person plural perspective so that selfhood becomes a matter also for the meaning of a 'we'. Inclusion as self-inclusion means thereby that the community is conditioned on the possibility to reaffirm the 'we' as a self or, more accurately, 'us' as *ourselves*. The concept of character shows the ease by which *ipse* turns into *idem*, lending the selfhood of the community qualitative attributes by requiring particular qualitative features of those who are making it up. When related to *idem*-identity, the selfhood of the community as a collective agent could be conditioned on how the people making it up share the quality of being, for example, Europeans, teachers, poor, tall, or green-eyed.

As individuals in the modern state, we are all presumed to be the same. Certainly, we are individually differentiated from each other, but we are this in the same way. We are presumed to have a selfhood and to be able to reflect on ourselves and be fully accountable for our actions, since we ourselves are the owners and agents of these actions.[2] Thus, the identity of the individual does not relax the problem of how the selfhood of the community is imbued with qualitative features even if these are reduced to numerical identity. Instead, the identity of the individual shows how numerical identity fosters internal unity by adding up a potentially unrestricted number of individuals that are recognised as such on the basis of a certain character. Sameness and selfhood have thereby become mutually implicating. For the self to be recognised as a *self* and thereby included in the community, a particular quality is required. Sameness and unity, in turn, are upheld by the continued reaffirmation of a 'we' that provides the community with selfhood, that is, the ability to reflect on itself and in doing so being able to say *we*, *us*, and *our*. The more accurate name for sameness would thus be self-sameness, since this indicates both a self that is the same in relation to itself and a collective sameness that demands the identity of the self. Thus, when transposed to the community, the distinction between *idem-* and *ipse*-identity cannot escape an image of the community as predicated on both sameness and difference.

Another approach to identity is to see how identity always already implicates difference. This is Jens Bartelson's conclusion when highlighting how "something can only be identical by virtue of being different from something else" (Bartelson 1998, 318). For identity to have a meaning, it needs to be understood in relation to difference. Two entities can only be distinguished as identical if they exist in a context of differences. Conversely, difference has only meaning "in a context of identities" (ibid.). However, if identity depends on the difference between two things, then those two things share the features of being both self-identical and different from each other. Therefore, in the sense of being identical to themselves while different from each other, the two things are identical also to each other. A blue vase is different from a red vase, but the blue vase has the identity of having a colour of its own which differs from the colour of the other vase. The same applies to the red vase. One could say that there is a sameness to the identity-difference relation and that this sameness, accordingly, presupposes both difference and identity and the relation between the two.

Bartelson addresses the identity-difference relation as mutually implicating, since the condition of possibility of identity is difference, and the condition of possibility of difference is identity. This means that nothing can be identical to itself only by virtue of itself. Instead, what is identical to itself is only so "by virtue of being different from itself" (ibid. 319). Identity thereby forms a "profoundly contradictory concept" as "the conditions of its possibility coincide with the conditions of its impossibility" (ibid.). Some approach this as a productive paradox and exploit the mechanisms of contestability at work in it (see Connolly 1991). Others use the paradox to scrutinise the tenacity by which certain concepts persist in their hold on political imagination even in critical discourse (see Bartelson 2001).

While Bartelson's conclusion is correct within the context of the conceptual analysis that forms his point of departure, the entire reasoning underpinning his argument rests on an idea *of difference as mediated by representation.*

It draws on how a concept, despite having a potentially infinite meaning, requires a determination when put to use. The reason that the paradox of identity and difference at all arises rests in how difference has been inscribed as *conceptual* difference. Now, conceptual difference invites the same problematic as does a paradox of identity that sees identity presuppose difference and difference identity. Grasped conceptually, difference must always fall back on its representation, which, if only temporary, forms an identity. While this makes Bartelson's argument regarding the difficulty in getting out of the trap between identity and difference very powerful, it also leaves his argument stuck in its own trap. Conceptually understood, difference must remain stuck in a hermetically closed relation, which puts its own mark of sameness on every transaction between difference and identity. If one instead seeks to escape this trap, then difference needs to be taken out of the setting where the problem first arises, namely the representational thinking that reduces the meaning of difference to the identity-difference relation.

Remaining within the realm of the conceptual, difference will always implicate identity also. To engage difference *without* negation, opposition, and contradiction would therefore require breaking loose from conceptual thinking (Deleuze 2004, xviii, 30). This would mean to intervene in the relation between difference and sameness and interrogate its effects. As Gilles Deleuze notices, difference understood as conceptual difference paves the way for resemblance so that the question "'What difference is there?' may always be transformed into: 'What resemblance is there?'" (ibid. 14). Emanating from the reflexivity of the concept of identity is a difference that doubles up with sameness so that being different to one another implies the possibility of being the same as one another. While this could ascribe to identity and difference the possibility to form stepping-stones for inclusion or belonging, it still fails to cut the link between the identity of the individual self and the sameness of the collective *we*—differences can be bridged and turned into resemblances, but this does not undo the problems emanating from a community predicated on a notion of sameness that presupposes difference. Hence, 'we' are still burdened by closures and exclusions.

Escaping a notion of community as predicated on sameness and difference appears to imply the near impossible task of withdrawing from representational thinking. This book reveals how sameness as a response to the question of community is complicit with language. It addresses how language, both in its referential and representational structure and in the distinction between those who have capacity to engage in language and those who do not, forms part of the problems of a community that assumes unity as its condition of possibility. Rather than a conceptual analysis, the book employs a hermeneutical deconstructive reading that studies the *relation* between the thinking of community and political existence and sameness, difference, and identity.

Variations regarding the substance of sameness and difference is of lesser importance than how sameness, difference, and identity are sustained by and productive of each other in various examples of the thinking of the community. For the critical ambition of the book, avoiding a definition of the meaning of community, sameness, and difference helps deepening the problematic of assuming an understanding of sameness that presupposes difference as the condition of possibility of community and sheds light on the implications this has for metaphysics, the political, emancipation, and coexistence. As concerns the normative ambition seen foremost in the elaborations on incalculable space, a hermeneutical deconstructive approach attends to texts by grasping how their meaning can lend thought to a community *not* predicated on sameness and difference as mutually implicating. Also here, the relation between sameness, difference, and identity is in focus, but now in terms of how these might escape being conditioned upon each other. Asking of a community that does not give rise to exclusions and otherness, it is suggested, would be to ask for *a notion of sameness that does not presuppose difference and a notion of difference that does not presuppose an assumption of identity.*

Why is it important to address the issue of sameness and seek to understand it without taking recourse to difference and identity? How we think about sameness has practical implications for politics and for communal belonging. When Hannah Arendt, more than half a century ago, engaged the community and the predicaments arising with it, she turned to the problem of statelessness:

> Much more stubborn in fact and much more far-reaching in consequence has been statelessness, the newest mass phenomenon in contemporary history, and the existence of an ever-growing new people comprised of stateless persons, the most symptomatic group in contemporary politics. (Arendt 1976, 276–77)

To Arendt, statelessness is a symptom of how a global political system of nation-states is productive of stateless persons since the right to migrate from one country does not correspond to a right to asylum in another (ibid. 280–82). I suggest that Arendt's stateless, if broadly understood as those lacking formal inclusion in the community of their presence, could be seen as symptomatic of the hold the notion of unity has on the thinking of the community and of the prevalence of sameness and difference lurking under its appearance. More exactly then, what is it the stateless as a group and phenomenon symptomatic of the sameness at work in contemporary politics is symptomatic of?

First, the stateless are a symptom of how the political in a community predicated on sameness and difference is structured along *boundaries*. Indeed, statelessness is symptomatic of how boundaries come to constitute *the very meaning of politics*. Boundaries cannot only be understood as the borders or limits of community but must also be grasped as boundaries to how communal belonging is imagined. To presume that borders only mark the outer boundaries of a spatial entity would be a mistake. While someone formally not included is inside the state in terms of residing on its territory, she is also on its outside, by being subjected to its power without constituting a subject of that power. In any case, she appears to be found on the wrong side of the border, making it difficult to answer if borders should be perceived as "limits on violence" or if claims should be made "about the violence of borders themselves" (Vaughan-Williams 2009, 60). That is, while granting inclusion to the formally excluded would undeniably help her particular situation, it would do less to relax the predicament of exclusion emanating from the boundaries of a community predicated on a notion of sameness that presupposes difference.

Boundaries mark out a particular *logos* of the community—they mark the limit of what can be compatible with and grasped by its rationale. Balibar sees this in the idealisation of borders and how they secure identities and particular views of the human. Marking out cultural difference, boundaries distinguish along the principle of proximity and distance when perceiving similarities and dissimilarities. Marking out civilizational difference, boundaries distinguish between the reconcilable and the irreconcilable, or the compatible and the incompatible (Balibar 2002, 94–95). While civilizational difference establishes the identity of the self against an other taken to be opposed to the self, cultural difference sees two identities form the outliers on a scale of possible identification. In both cases, identity in the form of a presumed internal similarity or sameness delineates the community and forms the trait by which the community is determined. Thus, any politics proceeding from identity when demarcating the community must be marked by an "intense receptivity to the existence of the Other" (Prozorov 2009, 27). Even if the sameness of identity only requires of beings that they are in possession of a personal identification number, it still serves to separate between beings and mark out the same from the different. Thus, if the stateless are a symptom of a politics structured around boundaries, it means they are symptomatic of a variety of manifestations of a notion of sameness that presupposes difference. These manifestations function to individualise, or even dehumanise, beings in accordance with the limits of the *logos* of the political community in question.

Second, the stateless are symptomatic of how being part of the community lets communal belonging take the form of membership. The assumed unity of the community suggests that with the community follows a question of membership, which, in the modern state, finds its form in citizenship. But with the community organised around notions of sameness that presupposes difference, such membership will be conditioned on one's alignment with whatever parameters of such sameness, be they the belonging to a particular minority or culture or the sharing in a particular *logos*. Hence, the stateless are symptomatic of how the 'we' making up a community predicated on sameness and difference calls for both the willingness and the ability of those aspiring for inclusion to conform to whatever the standards of similarity. Seeing belonging as a question of membership risks privileging an assimilation that requires the individual to conform to a given identity, instead of an integration directed to the reciprocal change of the new part of the community and the community alike.

Understanding communal belonging in terms of membership implies a failure in registering the actions of those lacking inclusion as political. The non-included appears instead only in her absence, since a particular identity is required of her to be registered as a political being. It therefore appears impossible for those not included to exist as political subjects in their own right, since the prerequisites for political subjectification are vested within a politics of inclusion beyond the control of the individual. This means that the formally non-included are caught in a relationship of fundamental inequality vis-à-vis the ones formally included in the community. When equal possibility to constitute a political subject comes in the form of an act of sovereign giving, the relationship of inequality at the boundaries of the community is once more reinstated. Equal recognition is then made "a *reward* rather than a *right*" (Shotter 1993, 128), due to it being received after the qualifications for membership are made.

Establishing communal belonging in terms of membership, a community predicated on a notion of sameness that presupposes difference implies a prior ascription to a particular category or representation for anyone seeking inclusion. If not fitting into any such categorisation or representation, the one without formal inclusion will remain outside the scope of the very process of acquiring membership. The question of membership further contaminates the ideals of democracy, since it concerns who is to be counted as a subject of democracy and hence as having a political voice that needs to be counted upon. If membership is what posits the measure for such a count, and if membership is grounded in a notion of sameness that presupposes difference, it follows that the normative ideals of democracy must take a detour through the relation between sameness and difference on which citizenship is based in order to be in effect. Consequently, the democratic ideals of participation, equality, and civil and political rights and liberties are all circumscribed by identity requirements established by community belonging as an issue of membership.

Finally, if the stateless are symptomatic of a community predicated on a notion of sameness that presupposes difference, then they are symptomatic of a closure of possibilities for what it means to be political as well as for the very meaning of politics. The criteria for the assumed sameness of the community lie outside the individual, in for example the nation, the group, a particular notion of humanity, or, for that matter, in a specific understanding of what procedures for engagement are considered as valid in a democracy. The same goes for political identity; a political identity that rests in an understanding of sameness established against what is different constitutes a closure of a multiplicity of possibilities and forms for political existence since the homogeneity of the particular will remain reductionist in nature. Here, the problem of the lacking attentiveness of the community to the politics of the non-included transpires also to the included, since a notion of sameness that presupposes difference posits a limit to what politics can amount to also for the ones formally part of the community.

Hence, sameness is a problem of how bounds are set to the meaning of politics and how political beings must be *made* recognisable, representable, and countable. Thus, sameness poses a fault line to what is irreconcilable with the community. If to be *in* requires being to some degree the *same*, sameness will draw a line in the sand against those who are just *too* different to be part. A community predicated on a notion of sameness that presupposes difference will therefore strive to ascertain that what is irreconcilable with the parameters on which its count of inclusion rests continues to be kept out.

In the discussion on *idem*- and *ipse*-identity, it was seen how also the identity of the individual was conditioned on as well as productive of sameness. This means that the problem of a community predicated on a sameness that presupposes difference is not only a problem for the nation, or for an acclaimed universal ethics, but also for the modern state. Thereby, the problem identified with Arendt emerges as related to the problem of representational thinking. Being an individualising power, the modern state is not developed above individuals, according to Michel Foucault. Quite the contrary, the condition for the integration of individuals in the modern state was that their "individuality would be shaped in a new form and submitted to a set of very specific patterns" (Foucault 1982, 783). The power of the state is such that it categorises the individual by attaching him to his own identity and turns this into a "law of truth". Whatever would separate the individual from himself is therefore concealed and constrained. Instead, his identity or individuality is something that "he himself must recognize and which others have to recognize in him" (ibid. 781). This means that modern state power turns individuals into subjects, in the sense that they are both "subject to someone else by control and dependence" and subject to their "own identity by a conscience or self-knowledge" (ibid.). Hence, being subjects of the state, individuals are not only subjugated to state power, but they are also subjugated to a rationale of an identity founded on an assumed self-knowledge and truth. Through this, state power is both an individualising and a totalising power (ibid. 781–82). On the one hand, it is a power that creates individualised subjects and does so on the grounds not of a particular identity but on the ground of *identity itself*, that is, on the assumption of there being a truth to identity—of there being a sameness to the self and an unambiguous answer to the Cartesian question of "Who am I?" On the other hand, it is a totalising power precisely by organising particular lives according to identity and, consequently, by structuring the population in conformance with the sameness of subjects individualised on the basis of identity.

This reveals something deeper concerning the problem of the stateless as a symptomatic group in contemporary politics. A symptom is something that gives clues about an illness, something that indicates a condition of disease and perhaps allows for a diagnosis. Such is the function of the stateless above, where it indicated the problem of how the closure of the community is established through a notion of sameness that presupposes difference. But the fact that statelessness is perceived as a presumably unwelcome symptom is perhaps a symptom in itself. If statelessness is symptomatic to the problems of what *we* are today—that is, what the very meaning of 'we' has come to entail—then statelessness as a symptom is perhaps symptomatic of how "the target nowadays is not to discover what we are but to refuse what we are" (ibid. 785). Thus, perceiving the symptom of statelessness as a symptom in itself does not warrant the search for a deeper and more refined answer to the meaning of the boundaries of 'us' and the question of how these can be made more inclusive. Instead, the reason the symptom of statelessness is symptomatic is precisely due to the hold a delineated 'us' has on the imagination of the community and the way by which the question of the political community is posed. This problem is twofold and played out both with respect to the community, which assumes sameness, unity, and identity as its condition of possibility, and with respect to the identity of the individuals constitutive of this community. If the meaning of 'we' emanates from an assumption of the political community as conditioned on a notion of sameness presupposing difference, then statelessness will always remain as a symptom of the ills of such a community. To Foucault, "the philosophical problem of our days is not to try to liberate the individual from the state and from the state's institutions but to liberate *us* both from the state and from the type of individualization which is linked to the state" (ibid., my emphasis). As it appears so far, to liberate *us* from the state and from the individualisations of the state would amount to liberating *us* from the boundaries that sameness imposes on *us* and that identity imposes on the individual. To ask what seeing statelessness *as a symptom* is symptomatic of means to assume another perspective on the question of the political community. It means to ask, why *stateless* and not *statefree*?

If statelessness is a symptom, it seems to be so because of a rationale that gives priority to the state. What would such a rationale that gives priority to the state consist in? Derrida describes it as a rationale vested in calculation. Through turning to Husserl and his discussion on the crisis of the European sciences as well as of European humanity, Derrida questions a rationality, a *ratio*, of calculability. While calculability is concerned with exactitude and objectivism, this is not the only mode by which reason can be conceived, Derrida maintains. As calculable, reason assumes a mastering authority of a particular world, which both brings the world into being and performs a closure on the world. Calculable rationality establishes a "mastering authority of architectonics" through which a certain world is set up according to "the unity of a regulative Idea of the world that authorizes that world in advance" (Derrida 2005, 121). This both unifying and appropriating reason will neutralise all that falls outside the scope of which it can give an account. In this sense, calculable reason exhibits an irrationality since it closes itself off to what it cannot account for. Derrida identifies a sovereignty to this reason that "knows everything and lets everything be known, that produces becoming or genesis but does not itself *become*, remaining withdrawn in an exemplary, hyperbolic fashion from becoming or from genesis" (ibid. 138). While finding this sovereignty of reason in Plato, Derrida sees it not only linked to the *politeia* but as proving a rationalism of the state; it is a state rationalism. Still, this closuring reason does not cause Derrida to give up on reason altogether. What he instead opens towards is a reason heterogeneous to the calculability of a *ratio*. Indeed, such a reason would mean to save reason from the irrationality inherent to its closure towards the unforeseen and the incalculable (ibid. 119–21, 132–33, 138–39).

If seen as resting on a rationality of calculability, a community predicated on a notion of sameness that presupposes difference establishes a particular count that determines both what counts as included and what at all is countable. Approached this way, it becomes clear how it is not enough to propose a different count of who makes up the community in order to do away with externalities. This is partly due to how the potential infinitude of numerical identity pairs up with qualitative identity, leaving also a more inclusive count stumbling against the boundaries of sameness and difference. What emerges is a community that is calculable according to a particular count that counts beings as included on the grounds of ascribing to a particular relation between sameness and difference. Being counted describes not only being summed up but implies also the possibility of being a countable being; of being someone that can be counted upon. Hence, counting is carried out on the basis of a measure that forms a closure since it separates beings that can be counted and counted upon from those who are uncountable. This indicates that a community predicated on sameness follows a rationale imbued with calculability. Only by being conceived according to the measure of such calculability can a being be grasped as countable and as a potential part of the community. The being in question conforms *already* to a particular order, since it is ordered or made orderable according to a particular count and a particular rationale.

Hence, the question of what statelessness is a symptom of appears to incriminate not only the identity of the subject and the state, but also a tenacious rationality of calculability. In other words, the unified community and the unitary subject are both accounted for by a reason vested in certitude. When posed with the assumption of unity as the condition of possibility of the community, the question of the political community will strive to render the community with certainty and exactitude. As will be further explored in this book, such certitude will delimit the meaning of the political and make the community ridden with exclusions and inequalities whilst deviating from democratic ideals. Assuming a notion of sameness that presupposes difference as a precondition for the political coexistence means to understand the community as anchored in calculability and kept within bounds, both as regards its subjects and its possibilities.

In contrast, Heidegger suggests that the political community is "that realm and locale around which everything question-worthy and uncanny turns in an exceptional sense" (Heidegger 1996b, 81). This suggests releasing the question of the political community from being posed within a *logos* of calculability. It suggests drifting the uncertain waters where conceptual thinking comes to a halt. The question of a notion of sameness that does not presuppose difference and a notion of difference that does not presuppose identity seems then ultimately to be *a question of an incalculable community.*

II

While this book addresses problems of crucial importance to real-world politics and that are felt in the everyday life especially of those not taken to be part of a particular community, its argument proceeds from theoretical sources and unfolds around the question of how Western *thinking* has assumed sameness as a necessary feature of the community. *De facto*, the sameness of the community is of course not that easily discerned, which forms part of the problem of assuming unity as crucial for the community. Sameness will be engaged in different ways throughout the book. The chapters explore the paradoxical democratic legitimisation of the people; attempts to broaden the scope for inclusion within various strands of citizenship theory; the role of boundaries and sameness for the continuation of politics; how a certain projection of space plays part in the formation and upholding of a unified people; and the persistence of the problem of calculability and incalculability and thus also of sameness. It could undoubtedly be seen as unfair to place texts on, for example, citizenship theory, which seek to present an empirical problem with empirical solutions, alongside more abstract philosophical explorations on the issue of a community predicated on sameness. Would it not be much to ask of those seeking to present on-the-ground-solutions to an acute problem that they also respond to the philosophical premises of this problem, one might duly ask?

It should, however, be noted that the failure to relinquish a community predicated on a notion of sameness that presupposes difference amongst some of the theorists is not necessarily a reason to abandon their attempts altogether. This book reveals instead how, despite efforts to reduce the effects of sameness, difference, and identity, the assumption of sameness and difference as precondition for the community has very persistent effects, leaving also emancipatory claims questionable. Guiding the reading of various authors are the questions of: What is the relation between sameness and difference in the texts? What is the relation between difference and identity? And how do the texts pose or withdraw from a calculable community? On the one part, the aim of the reading is to critically assess the responses to these issues. On the other part, it seeks to hermeneutically understand what the texts allow us to think with respect to a community *not* predicated on a notion of sameness that presupposes difference and what they allow us to grasp of the meaning of such a community. For that, their failures are as good a help as their moments of success.

The critical approach of the book seeks to grasp the role the assumption of unity as a condition of possibility of the community has in sustaining the relation between a perceived whole of the community and those who form an other part than described by the community's self-perception. Rather than revealing an understanding of sameness that presupposes difference as something the question of the political community is irreducibly locked up in, the import of insights gained is to reveal the implications of a community predicated on sameness, difference, and identity for how politics, inclusion, boundaries, and political existence are understood.

The normative and hermeneutical approach of the book is, perhaps somewhat puzzlingly, most apparent in its reading of the most dubious author and the most problematic text, namely Martin Heidegger and his lecture on Hölderlin's hymnal poetry on the Ister River. To read Heidegger to explore the meaning of a community *not* predicated on sameness, difference, and identity is a precarious move that might certainly appear as counterintuitive. To read a text like *The Ister* for the same purpose—a text based on a lecture given in 1942, a text that holds explicit references to National Socialism, and in which Heidegger attempts to give a new meaning to the German people through engaging Friedrich Hölderlin, a poet embraced by nationalist thinking—might appear close to foolish.

What Heidegger helps open up is a thinking that does not, or at least claims to not, rest on a particular ground. Heidegger engages an existence that is ultimately *not* grounded in an origin, a place, or a time. Instead, he seeks to think the groundless, both individually and collectively, and turns thereby *thinking* into a limited experience, as thinking the groundless forms an attempt to think something that does not let itself be thought.

For Heidegger, thinking takes off with the insight of the world as marvellous and incredible, as uncanny and strange. The thought needed to grasp such a world can never be found as long as the world remains inscribed in a systematising grid, since such a grid would mean that the world can be read only in a particular and hence already restricted way. Rather, thought must open itself onto the world—and onto itself. Thinking, then, is a transgressive undertaking; it transcends what is immediately presented and dives into the voids opened up by its own activity. Thinking opens up and clears a space, but not in the sense of claiming a delineated area for a particular kind of thought. Rather, thinking "is a clearing" in the sense that it "allows for a new light to shine. Whenever thinking happens, beings come to shine anew" (de Beistegui 1998, 1). This means that thinking also opens towards the future, or, as Miguel de Beistegui puts it, "thinking is itself futural: it is a leap ahead, a transgression that opens onto another present, another historical possibility, a 'dangerous perhaps'" (ibid.).

Why a *dangerous* perhaps? Because all thinking involves a risk to the extent that: "To think is to risk" (ibid.). Inherent to every thought lies a potential risk both of uncovering the essential groundlessness and nothingness of thought and of closing that space that has been cleared through thinking. Thinking forms "essentially a departure: de-parting from the present, thinking ventures into the promise of another time, another configuration" (ibid. 1–2).

For Heidegger, the work of art plays a crucial role in such a de-parting. Since art exhibits the special quality of being poetic, it makes a basis for a thinking that is a *departure*. Poetry stands out as the crucial source for thinking to Heidegger as it allows one to approach the world in a poetic mode. Heidegger plays with the German word *dichten*, which, alongside meaning "to poetise", also means to condense or concentrate. A work of art has the potential to give a thicker, denser, and more multifaceted rendition of the world (Ruin 2013, 130). Through the work of art, through poetic writing, a more complex and versatile image of the world can be drawn. Such an image is not immediately accessible in all its aspects and does not lend itself to any quick and uniform answers. What art instead does is to allow for venues of thinking to unfold. The work of art has the ability to move its spectator into that space that needs to be cleared in order for thinking to depart from the present and venture into the new and unknown.

It was during the first half of the 1930s that Heidegger turned to the intersection of art or poetry and philosophy. This was also when Heidegger's involvement with the National Socialists was at its height. Soon, Friedrich Hölderlin stood out as the poet with whom Heidegger especially sought a dialogue. Heidegger held his first lecture on Hölderlin soon after he resigned as rector in Freiburg. The last was given in 1942, that is, three years before he terminated his membership in the NSDAP.[3] Heidegger's fascination with Hölderlin unfolds alongside the interest in the poet shown by the Nazi regime, which appropriated Hölderlin, and also Nietzsche, for the Nazi propaganda and historical narrative, making them part of the "mythologized pantheon of Nazism" (Elden 2001, 38). Hölderlin was especially appropriated as a poet of the German and seen as "the purest of all German poets and the most purely German of all poets" (Savage 2008, 8). The reception of Hölderlin sees the poet emerge in two forms—one in which he is canonised and appropriated for edification in a particular time, another in which he remains "susceptible to pain and gifted with vision of an almost hallucinatory clarity", or, put differently, one Hölderlin "as he is judged by Germany, the other as he judges Germany" (ibid. 24). It is the latter Hölderlin to whom Heidegger was drawn.

Although Heidegger's Hölderlin lectures make explicit references to National Socialism, and at times share a contemporary discourse on homeland, people, and soil, they also serve to distance Heidegger from the racist nationalist politics of his time. Heidegger's engagement with Nazism has been afforded numerous volumes of literature, and it should not be doubted that Heidegger stood, at least during a few years, very certain behind the Nazi movement. As Julian Young puts it, "His involvement was a matter of conviction rather than compromise, opportunism, or cowardice" (Young 1997, 4). Of central concern, for the ones engaging with the "Heidegger controversy", is whether his complicity with Nazism is a consequence of his philosophy. Although Heidegger might have been convinced that Nazism would function as an extension of his philosophy, the question is if fascist politics is the logical outcome of Heidegger's thought. Young disputes this to be the case and emphasises that Heidegger's rejection of Nazism in 1945 should be understood as a serious dissociation, something that the publications of his so-called Black Notebooks might counter. However, Young argues for the inconsistency between Heidegger's political involvement and the deeper tenors of the philosophy he had already worked out at the time when the National Socialists came to power in Germany (ibid.).

Still, in April of 1933, Heidegger was elected as the first university principal under the Nazi regime. Although he decided to leave the position as principal within less than a year, the fact that he assumed it in the first place meant he took stance for National Socialism and rendered the movement certain legitimacy. If one seeks to understand Heidegger's relation to and expectations of the National Socialist politics and National Socialist revolution, the primary text to turn to is the rectoral address he gave in May 1933 when he shouldered the duties of the Rectorship of the University of Freiburg (Ruin 2013, 151–57). The address circled around the question of the role, essence, and organisation of the German university and was saturated with allusions to leadership [*Führung*]. It is a both philosophical and political work (de Beistegui 1998). The address is philosophical in the way it centres on the ability *to question*. It is political, not only in the obvious terms of being held by an employee of the National Socialist state and constituting an endorsement of that state, but also through identifying the political as "the *repetition* of an historical possibility, that of questioning", which in extension means the possibility of revolution (ibid.). Politics here becomes subordinated to philosophy, and both de Beistegui and Hans Ruin urge to not simply dismiss the address but rather to turn to its philosophical content and see how it plays into the politics it expresses and takes part in (ibid. 35; Ruin 2013, 157–60). Questioning shatters, breaks open, "blows the scientific landscape into pieces", but it also gathers and grounds and attunes the forces set free (de Beistegui 1998, 52–53).

However, Heidegger does not direct this account of questioning, which lies at the core of the university, towards the gruesome political implications of his own speech. The 'we' that Heidegger articulates with such ease in the rectoral address is not a 'we' pronounced with any discreet subtleness of academic discourse, but rather a highly political 'we'. Maybe, as de Beistegui says, politics begins when an 'I' speaks in the name of a 'we' (ibid. 45). In contrast to *Being and Time*, which refrained from the plural and rather described man as singular essence, the Heidegger of 1933 turned to a very particular we—that of the German people; the *Volk* and the *Volksgemeinschaft* (ibid. 45–46). The rhetoric is in unison with the time in which it was uttered, as Heidegger articulates the spiritual world of a *Volk* to be neither its "cultural superstructure" nor its "arsenal of useful knowledge", but rather to be "the power that comes from preserving the forces that are rooted in the soil and blood of a *Volk*, the power to arouse most inwardly and to shake most extensively the *Volk*'s existence" (Heidegger, the rectoral address, quoted in de Beistegui 1998, 54).

While in the rectoral address the 'we', the national, is marred with allusions to a highly particular German *Volk*, the 'we' in Heidegger's reading of Hölderlin appears in a different light. There, Heidegger approaches Hölderlin with the aim to open up the 'we', to phrase the 'we' as a question rather than a fact. While Heidegger's engagement with Nietzsche sought to wrest Nietzsche out of the biologist Nazi representation of the philosopher, the readings of Hölderlin remain within the question of the national (de Beistegui 1998; Elden 2001, 38–39; Ruin 2013, 133–34). Still, this does not correspond to any kind of straightforward nationalism, but rather the national itself is put in question (Ruin 2013, 133–34). Heidegger pursues his conversation[4] with Hölderlin in order to uncover the national by questioning the national in itself (ibid.). Thus, Heidegger approaches Hölderlin partly to understand a national that is not yet sealed in a particular meaning. "At stake", de Beistegui notices, "is the possibility to think the national 'before' any decision regarding the nation-state and independently of the question regarding the juridical status of nationality" (de Beistegui 1998, 88). Nevertheless, the Hölderlin lectures constitute no move away from the national itself. While Heidegger unambiguously criticised biologist and racist contemporary nationalism, he still expressed a cultural and linguistic nationalism himself (Elden 1999, 265). In addition, the Hölderlin lectures are imbued with a heavily burdened language and circle around the homeland, homelessness, homecoming, and the foreign, while they also address the shortcomings and possibilities of the National Socialist movement. Considering his attentiveness to language, it would be erroneous to assume that Heidegger was unaware of the implications of the language he engages. Although he asserts that terms such as "homecoming" should not be approached through a patriotic or nationalist lens, Hölderlin is used not only to think through a philosophical problem; there is also a political project to the engagement with Hölderlin—to make the poet "a power in the history of our people" (Heidegger, quoted in Elden 1999, 264). Thus, Heidegger "is nowhere as close to philosophising in the service of the nation and the national as in some of the Hölderlin lectures" (Ruin 2013, 146, my translation). So why at all turn to Heidegger and a text such as *The Ister* in order to ask for a notion of sameness that does not presuppose difference? What can a poem about a river tell of an incalculable community, especially if this river seems so embedded in a nationalist discourse and imaginary?

When Heidegger thinks of the groundless in relation to the community in *The Ister*, he draws a community that is not constituted through its roots but in relation to the possibility of thinking the foreign. If thinking strives towards the groundless, if it seeks to avoid any closure of the space opened up by thought, then the foreign cannot be thought of as something static or determined. At stake, then, in *The Ister* is to think the foreign without *securing* the foreign. To the extent that thinking forms a move towards the open and the groundless, then *The Ister* also forms an attempt to move away from the calculative thinking that seeks to master the world by fixating it in rigid categories and ordered measurement. In pre-Socratic Greek thought, Heidegger locates a foreign that escapes any securing. He returns frequently to the correspondence between Hölderlin and Böhlendorf in which Hölderlin writes of what is natural to the Greeks as "the fire from the heaven", while what is original to the Germans is the "clarity of representation".

To Hölderlin, the clarity of representation involves the scaffolding, the frames, the closures and compartments used for the presentation and constitution of the conceivable (de Beistegui 1998, 111). For Heidegger, this forms part of the predicament of an *orderly thinking*. Modern thinking operates so that it organises the world in terms of order, in an orderly thinking. This is a calculative thinking that allows human life to be "condensed into diminutive numerical values for the purposes of our calculative planning far in advance" (Heidegger 1996b, 39). The Greeks forms a reversal of this as they stand in direct proximity to the inconceivable. For Heidegger, the German 'we' must rid itself of the clarity of presentation and learn from the encounter with the foreign pre-Socratic Greek. Hölderlin's poetry is assumed to open a new historical grounding, indeed, to have "*already grounded* another possibility" in forwarding a unique historical relationship between the German and the Greek (Janicaud 1993, 388). In the Ister lecture, the poetic stream is assumed to determine "the coming back home of the Western man" (ibid.). At stake here is a historical grounding deviating from a historiography that writes its temporality through entities, facts, and concepts. Hölderlin's historical grounding is instead characterised by a historicity, a *Geschichtlichkeit* through which historical unfolding is vested in a questioning encounter between the German and the pre-Socratic Greek.

Given its nationalist undertones, its emphasis on the foreignness of the Greek, and its critique of calculative thinking, there are different ways of reading *The Ister*. One can see it as an imperialist text both in the sense of a privilege given to a history of the West framed through the Greek and the German (ibid. 110) and through the lacking elaboration of how Hölderlin—taken to articulate a truth on the question of being—is placed in a particular people and culture (Ruin 2013, 147). To reencounter their lost essence, Heidegger asserts that the German people need to turn to what is most alien and foreign to them, be it the long-lost time of Ancient Greece or the other of the colony. By forgetting themselves in their encounter with a foreign time or place, a new path towards the essence of their own might be uncovered. One can also read the Ister lecture as a book about loss and longing for a vanished Greek thinking. Such a nostalgic gaze at a previously enchanted cosmos resonates with what Angharad Closs Stephens calls nationalist time. A nationalist temporality experiences time as linear and continuous; what once was can again come to presence. Nationalist time draws a univocal and linear history reflected in the continuity and homogeneity of the people. This means that politics becomes a matter of mastering time and serves as a unifying promise creating stability, order, and continuity. Hence, a move away from the idea of a unified nation must also involve a rejection of a unifying account of time and politics (Stephens 2013). It is therefore important to bear in mind the narratives of loss and longing at work in Heidegger's lecture. It is also critical to be aware of the imperialist tendencies of the text, which makes *The Ister* vulnerable to the critique of how "the possibility of advancement in history is *only* available to those people that are *already* bound within a common ... political unit" (Stephens 2010).

As Janicaud points out, following Hölderlin's poetic river, Heidegger's *one law* of history forms a founding that inexorably brings him back to a grounding and a metaphysics, though a metaphysics that has abandoned many of Heidegger's earlier presuppositions. It is also a history that radically places man in the West and, indeed, names man as the Western man. Still, Janicaud does not suggest abandoning Heidegger's engagement with Hölderlin altogether but encourages instead the reader to approach Heidegger's engagement with Hölderlin without thinking "that there is *one* law of history, but rather that our history is still open, still full of undecidable voyages, downstream, upstream" (Janicaud 1993, 388–91, at 391). The latter suggestion guides the reading of Heidegger and Hölderlin's "The Ister" undertaken in this book. My reading is both close and somewhat unfaithful to Heidegger's text. It seeks to explore a notion of space that is not bound up in a notion of sameness that presupposes difference. Whereas Heidegger is commonly engaged as a philosopher of time, history, and language, this book follows a more narrow body of thought that reads him as a philosopher of space.[5] It is suggested that attending to the relation between space and time, and especially the space of the community, helps lend thought to a community not predicated on a notion of sameness that presupposes difference. I argue that from Heidegger's understanding of space or place in *The Ister* and some other texts follows a sense of community that escapes being delineated through number or semblance. If the open, fluent, and shifting space of the river is to be sustained, then the community cannot be grasped in terms of sameness. Instead, the space that unfolds from the river withdraws from the "logic of territorialisation in favour of a dwelling that is essentially homeless" (de Beistegui 1998, 178n4). This is a space that shuns the thought of a community of blood and soil and instead exposes "the homelessness of the human condition" (ibid.).

Engaging Western political *thinking* of the community, this book is largely theoretical. However, it returns to the question of the stateless and touches also on the example of the case of Papperslösa Stockholm, a network of undocumented migrants working for the regularisation of undocumented migrants in Sweden. Their claims concern a right to health care and labour rights. Working without a working permit is illegal in Sweden, and the network argues that the regulations of the labour market should be extended to everyone, including undocumented persons. Reoccurring in the book is also facets from Sophocles' drama *Antigone*. For several authors studied, the *Antigone* tragedy is engaged to say something about the community, politics, and *polis*. In brief, the tragedy unfolds around the royal household of Thebes and especially the four children of Oedipus and Jocasta. Their two sons, Eteocles and Polyneices, die by each other's hand in a fight over the *polis* of Thebes, with Eteocles defending the newly established democracy and Polyneices challenging it. Creon, the king of the *polis* and the maternal uncle of the children, declares that Polyneices should be left neither buried nor mourned since he defied the laws of the *polis*. The opening scene introduces the two royal daughters, Antigone and Ismene, discussing what is to be done with their brother who is left to rot outside the city walls. Deciding to defy Creon's decree, Antigone buries her dead brother, loudly lamenting her loss according to ritual. She is discovered in the act and condemned to be entombed alive despite protests from her betrothed, Haemon, Creon's son. Eventually convinced of the gods' disagreement of his act, Creon decides to bury Polyneices' ravaged body and let Antigone free from her tomb. Nonetheless, his arrival to the tomb comes too late and Antigone has already hanged herself, making Haemon take his life in grief, an act repeated by his mother, Creon's wife, Eurydice. The final scene sees Creon led off stage lamenting his fate as no one has heard his call to put him out of his misery (as narrated by Elden 2013a; Honig 2013).

The study conducted here aligns much with research contesting the hold nationalist sentiments have on political imagination. Much of this research has focused on the temporality of the community and how this furthers nationalism (Anderson 2006; Bhabha 1990; Edensor 2006; Shapiro 2000; Stephens 2010, 2013). Stephens finds attempts to escape nationalist understandings of the political community often imbued with a nationalist rationale due to their reliance on linear history and a narrative of loss of a once vibrant origin (Stephens 2013). While touching upon the role a linear and teleological time plays for sustaining the assumed unity of the community, the dominant focus of this book is on space and place. Thereby it aligns with engagements with space within political and human geography. Some of these have focused particularly on identity and relational space (Massey 2005; Scott 2009). Other have brought to attention the space of the city and its potential to rethink identity and the meaning of the political (Coward 2008; Derrida 2001; Isin 2002, 2007; McFarlane 2011; Stephens 2010, 2013), while political theory has explored theoretical and philosophical implications of engaging with spatiality (Elden 2001; Malpas 2012; Shindo 2012; Sloterdijk 2011). Contrary to some of these examples, the study pursued here discusses less the meaning of spatiality for identity formation. By contrast, it seeks to elaborate a notion of place as something that *escapes* identity.

Research on nationalism and community has focused on the moral psychology of nationalism to show the different meanings the concept of the nation has attained throughout history (Yack 2012) or to attempt to theorise states *without* nations through studying the laws of kinship at work in hereditary groups (Stevens 2010). Alongside targeting nationalist thinking, the question of the political community is here directed also towards the state and how the state is crucial for upholding a notion of sameness that presupposes difference as the assumed condition of possibility for the community.

Engagements with the problem of how to broaden the scope for inclusion to make place for those other to the community take place either within a body of thought that centres on the question of the *demos* and its legitimacy (Abizadeh 2008; Bohman 2010; Canovan 2005; Espejo 2011; Näsström 2007) or amongst discussions on citizenship and its limits (Benhabib 2004; Erskine 2000; Young 2011; Yuval-Davis 1997b). Only a few of these hint at the assumed unity of the community as productive of the problem of delimitation and exclusion (Näsström 2010). This is instead what the agonist political theory discussed in chapter 2 concentrates on (Connolly 1991; Honig 2001). Nevertheless, it is argued that these examples also stumble upon a community predicated on an understanding of sameness that presupposes difference, although some steps are made to shift the arena for political action and contestation.

The question of the unity of the community is also engaged within international relations theory, where it is often articulated as a dialectics between the universal and the particular. The deadlock between the two and its effects on the inside and outside of community is well elucidated by R. B. J. Walker, who underscores that any universalist claims are bound to be particular, since they are saturated with an array of boundaries, limits, and borders. Thus, attempts to reimagine political possibilities are wise to defy universalising tendencies since these tend to assume a particularist and state-centrist outlook (Walker 2010, 184–258). Bartelson traces historically the meaning of the concept of world community and shows how our cosmological views ground our conceptions of community and, conversely, how conceptions of community serve to make sense of a common human cosmology. Here, he problematises the trap of differential identity the previous section saw his reasoning caught up in. If a common mankind is articulated through a human identity presupposing difference, then any notion of community is likely to be marked by the same problem (Bartelson 2009b, 11–12). Bartelson seeks a resolution to this quandary in human sociability (ibid. 178). Similarly, Stephens finds a remedy to nationalist time in the place of the city, which allows for Jean-Luc Nancy's notion of *encounter*. With this, she gestures towards positions not secured by identity but instead co-determined through the mutual interaction at work in encounters and how the city forms a place inviting these. Stephens notices that to describe the community as other than a thing involves engaging "the difference between community as something that can be calculated and community as something that escapes calculation" (Stephens 2010, 39) without following the path of incalculability much further. Discussing in part the same theorists as Stephens, this study also examines the notion of encounter. Yet it takes a different position concerning the place of encounter than do Stephens and Bartelson, who both must rely on a certain openness, visibility, and capacity for subjects to engage directly with one another, a capacity or liberty this books contests that those most vulnerable to exclusions might not be able to exercise.

Sergei Prozorov's elaborations on universalism (Prozorov 2014a, 2014b) counter both Walker and Bartelson and posit instead a universalism that is *void* and that allows us to speak of the World neither as something, nor as everything, but as *nothing*. Drawing on continental philosophy and especially Badiou, Agamben, and Heidegger, Prozorov contends that life unfolds in a vast plurality of particular positive worlds, ontologically grounded in the universality of the World. This universal World, which each singular world holds something of, is nothing. Hence, what appears as something in the singular positive world is ontologically nothing when approached through the universal World. Conversely, the nothing of the universal World will always emerge as something in the positive world of its appearance (Prozorov 2014a, 27–35). This leads Prozorov to revitalise the well-known triad of community, freedom, and equality as the three axioms of politics. All three axioms emanate from the void of the World, meaning that all beings who are in the World are so as free, equal, and in common. Since the World is nothing, it lacks determinations, leaving beings free also from identity and thus from themselves while simultaneously exposing the nothingness of a particular world (ibid. 79–87).

Although approaching the problem of universalism with algebraic precision, Prozorov nevertheless fails in thinking a community of the world that does not presuppose difference. The contours of a community not predicated on sameness emerge at the moments when political action contests the limitations of a positive world. Still, the possibilities of such a community are pushed to the ontological *outside* of any world, leaving little hopes for a positive world to escape exclusions. Since the respective positive worlds are always *something*, it follows that *in* them the persistence of sameness, difference, and identity is left intact. My approach to the problem differs, by seeking to grasp the meaning of an incalculable community immanent to the world. A gradual relocation of the question of the community will be seen throughout the book, in that the question is moved from an above determination to instead emerge in the relations between beings. Hence, while Prozorov seeks to formulate an ontology of a Politics and World *for all* that nevertheless lies outside any positive world, this book attends instead to an immanent world where there is place for all.

III

The rest of this book is organised as follows. The first part dwells on attempts to expand the community. Chapter 1 begins with a discussion on the paradoxical democratic legitimacy of the people. On the one hand, the democratic people is supposed to emerge on the basis of the laws stipulating their constitution. On the other hand, for those laws to be legitimate, they must emanate from the people themselves, that is, from those whose legitimacy they should guarantee. This means that a chicken and egg problem lies at the heart of democratic theory since, although being the source of the legitimacy of authority, the people fails legitimising itself. The chapter argues that the reason the problem of a lacking legitimacy of the constitution of the people at all arises is found in the assumption of unity as a necessary prerequisite for the political community. The assumed necessity of unity as the condition of possibility of the political community has often invoked the nation as a stand-in for the lacking legitimacy of the people. The chapter proceeds to study how attempts within citizenship theory to expand the community are prone to exclusions. It is argued that grasping groups through *idem*-identity will result in an understanding of political belonging as a question of membership. It is further argued that, in embracing *idem*-identity and valorising the particular character of the collective, assimilation, rather than integration, becomes the central requirement for membership. However, it is also argued that embracing selfhood or *ipse*-identity as a ground for inclusion cannot fully fend off the risk of turning integration into assimilation. Instead, when perceived as a global pedagogical and ordering project, the category of citizenship emerges as an *idem*-identity in itself, with exclusions to follow.

Finally, the chapter engages what is grasped as a paradox of democracy. On the one hand, democracy is based in the numerous voices and opinions of the *demos*, that is, democracy is based on the incalculability of the *demos*. On the other hand, democracy depends on the numerous voices of its incalculable authority being counted equally. This, however, implies that a measure for whom to count and how to carry out that count is necessary. The question of political community, then, emerges as not only a question of membership but also as the double question of *who counts*.

Chapter 2 explores how agonist political theory embraces the paradox of politics as a means to keep the boundaries of the political community open to contestation and change. Democracy comes thereby to constitute a solution to a hegemonic collective identity. Still, since limits are what animates pluralisation and political contestation, both the meaning of politics and the question of the political community will be structured around boundaries. The chapter continues to unpack Rancière's understanding of democracy as a disjunctive junction between power and *demos*. Rancière locates the political in a disagreement of the boundaries of how the people is conceived. This disagreement gains its force from the enactment of an assumption of the equality of *logos*. On the one hand, this sees the non-formally included as the primary protagonists of political action and brings them to the centre of politics. On the other hand, in addition to the issue of how the expansion of community must be accompanied by closure for political contestation to be able to persist, Rancière's assumption of equality as a given also rests on a contemporary supposition of a certain degree of freedom. It is argued that if freedom is understood as living in the absence of dependence on the goodwill of others, then Rancière's assumption of equality as a given becomes circumscribed by liberty. As Rancière's theory of political enactment is located in the realm of speech, *logos*, and visibility, this implies that those in society who lead a clandestine existence and are the most vulnerable and most exposed to the inequalities stemming from an exclusive community are barred from the political community. Whereas agonist political theory, then, seeks to engage the question of political community in terms of how political contestation can be enacted, this must be accompanied by the question of by what capacity one is able to act and emerge as a political being in the first place.

Chapter 3 turns instead to the issue of space. It notices how boundaries are at work not only in whatever requirements are assumed necessary for belonging to the people, or in the particular *logos* that structures inclusion in the community, but also in the very meaning of the political and of political action. In addition, this chapter engages boundaries as boundaries of space. This is done both through unpacking the territorial border and the spatial assumptions this rests on and through scrutinising boundaries of space as limits to how we think of the spatiality of the community. The chapter explores the role of mapping, cartography, and geography for the emergence of bounded space. It is here seen how bounded, calculable space was crucial for the emergence of modern authority. Engaging the changing character of sovereign authority with respect to domestic control, it is maintained that space of modern authority is a space of sameness rather than territory. A space of sameness forms a spatiality that differentiates, fragmentises, and installs hierarchies in space according to a rationale of sameness that makes any form of difference manageable from the *logos* of the centre.

The chapter proceeds to study engagements with space that seek to release space from the spatiality of authority. Here, the relational character of spatial encounters is seen to gain prevalence over a spatiality based in representational identity. Likewise, interstitial spaces between signification and sign emerge as spaces of contestation, re-articulation, and manoeuvre, in which the post-colonial subject can alter her position vis-à-vis the state. Space in the form of unmapped terrain is also seen to be complicit in the escape from state identity. Yet, their fruitful engagement in releasing space from calculability notwithstanding, these attempts are all set up *against* a spatiality of representation, calculation, and conceptualisation. Hence, the spatialities they describe form mere aberrations to the bounded space of the community. Thus, the question of political community becomes a question of locating it rather than *relocation*. To relocate the political community, the chapter concludes, drawing on Arendt's engagement with place, means not to seek it *elsewhere*. Rather, it means to locate it anew, that is, to find another way of grasping its place. Thus, the next part of the book approaches the question of the political community through asking how the way we think about the place, space, and spatiality of the political community affects the ways we think about belonging, coexistence, and political action.

The next part of the book engages the notion of an incalculable community. Chapter 4 begins with an attempt to uncover the meaning of incalculable space and grasp the implications this has for the meaning of the political community. This is done through a close engagement with Heidegger's notion of dwelling and his reading of Hölderlin's poem "The Ister". First, however, Heidegger's own understanding of identity, sameness, and difference is explored. Here, it is seen how Heidegger seeks to avoid the metaphysical principle of identity, which assumes that two entities can be identical to each other and that turns the same and the different into subcategories of identity. For Heidegger, this forms merely one way of understanding how beings appear in the world. Instead, he seeks to understand how sameness and difference belong together, not through being structured under the category of identity, but through how they relate to each other while yet being held apart in the way they enter and make the world appear.

Spatiality forms a path out of the metaphysical principle of identity for Heidegger. This chapter takes up the notions of river space as the place of the community and dwelling space as the place of human existence. River space is elaborated as an ephemeral space that draws the place of the community as vested with flow, emergence, and withdrawal. With river space, the space of community emerges as both vanishing and under way and must always be understood as different in relation to itself. The chapter explores how the river gives way to a spatiality that does not run aground in a temporary sign, representation, or identity. Instead river space forms an incalculable space that emerges as *prior* to determination, as ahead of both *logos* and calculation. Heidegger's figure of dwelling is further engaged to understand how beings gather space through dwelling in it. Whereas Heidegger sees dwelling as a human activity, I suggest that anyone can dwell. With dwelling, place consists in an intimate relation with the world, where beings withdraw from calculability and representation, experience the world as felt, and form the world pressing upon it with their bodies. As the spatiality of place is also marked by the flow of the river, the chapter suggests that place describes a difference that does not presuppose identity. The chapter proceeds to see how the notions of the river and of dwelling release place, including the place of the political community, from delineations. It further attends to how the figure of dwelling locates the formally non-included as already part of the fabric of the community and as partaking in its emergence, without taking recourse to *logos*.

Drawing on the insights from the previous chapter, chapter 5 formulates a notion of sameness that does not presuppose difference as *a predicament of situation*. This describes the shared problem of being torn between *situation*, understood as the lived, manifold experience of being in place, and a *position* in which calculable reason orders beings according to categorisation. The chapter maintains that sameness as a predicament of situation does not give rise to a unified us. What is shared in the predicament of situation is rather a lack, stemming from the tension between how the rationality of modernity posits beings in calculable positions and by being in a situation of disclosure and presencing experienced through dwelling. Still, attending to the predicament of situation, it is suggested, implies a continued openness to how beings emerged in place form part of the disclosure of the community.

Although chapter 4 sought to broaden the meaning of dwelling, Heidegger's own use of the notion gives priority to the human. To better grasp what is entailed in the valorisation of the human, the chapter turns to Derrida's engagement with the contentious reader. It discusses how Heidegger's centrality of language eventually awarded the possibility to come to presence as integrated world-forming beings to a limited few. The chapter continues to study how Derrida and Nancy open up the meaning of language and make room for difference within language, something that in part is done through elaborations on the relation between calculable and incalculable reason. The final parts of the chapter discuss the limitations Derrida and Nancy encounter, due to how they give priority to reason, either as a measure of equality bound to delimit the possibilities of incalculable freedom or through relying on the capacity of calculable reason to expand its reach so as to make room for the incalculable. This relation between an incalculable freedom and a notion of equality based in calculable reason is another way to formulate the problem of the relation between an incalculable *demos* and a democratic authority that requires calculation. The concluding discussion of the book engages this problem through the meaning of sameness as a predicament of situation and place as a difference without identity. It returns to the issues of membership, territory, and the meaning of politics and discusses how these can be understood in the light of the incalculability of place that makes way for an incalculable community.

NOTES

1. Studying undocumented migrants in Sweden, Erika Sigvardsdotter shows how citizenship forms such a border. In its perhaps both most abstract and concrete form, citizenship provides a numerical interface between the resident and the state. For those lacking a personal identity number, anything that demands identification becomes close to impossible; from leasing a flat to being in contact with public and civic institutions (Sigvardsdotter 2012). This indicates the ubiquitous character of borders but also how a minimal degree of sameness separates included from excluded and establishes boundaries between people.

2. Already here exclusions are at hand. Children and persons with intellectual disabilities, for example, are not always considered to fully conform to the standards of individual selfhood. This counts also for women in a not all too distant past.

3. Heidegger's Hölderlin lectures were held from the late first half of the 1930s to the early 1940s. The course on the poems "Germania" and "The Rhine" was one of the first lectures given by Heidegger after his resignation from his position as rector. This was followed by a course on the "Remembrance" hymn in the winter semester of 1941–1942. The lecture based on the hymn central to this text, "The Ister", was given in the summer semester of 1942 in a course that also examined Sophocles' *Antigone*.

4. The German term *Auseinandersetzung* characterises Heidegger's reading of both Nietzsche and Hölderlin but also of his simultaneous confrontation with National Socialism. *Auseinandersetzung*, or *Aus-einander-setzung*, literally means setting apart from one another.

5. Despite being foremost associated with a philosophy of time, a number of scholars within various disciplines have attended to Heidegger as a philosopher of space, and especially the spatiality of the concepts of dwelling and place. Several authors have helped deepen the insights of spatial aspects of Heidegger's philosophy and argued for adopting a topographical approach when reading his work (Elden 1999, 2000, 2001; Malpas 2006, 2012; Schatzki 2007; Vallega 2003). Vallega suggests that Heidegger's spatiality elucidates how thinking unfolds on exilic grounds, that is, on grounds that never are one's own and only temporarily lend themselves as foundations (Vallega 2003). David Kolb emphasises the importance of engaging place rather than space, but contests the totality of a unified people he suggests emerges from Heidegger's place when paired with the notion of thrownness (Kolb 1990, 149–58; see also Harvey 1996, 299–316). The relation between place and dwelling has gained geographers' attention. Edward Relph draws in part on Heidegger when discussing an embedded and lived notion of place in contrast to modern placelessness (Relph 1976). Dwelling has further been deployed to explore the relation between landscape, community, and place and has been taken to describe an embodied rootedness in place (Cloke and Jones 2001). Anthropologist Tim Ingold, in turn, sees landscapes emerge intertwined with taskscapes when adopting a "dwelling perspective" that explores how to dwell is to inhabit. This sees humans as part of the continuous transformation of the world, not merely through acting upon it but through moving with it (Ingold 1993, 1995, 2005). Social theorists Phil Macnaghten and John Urry make use of Ingold's notion of "dwelling perspective" when rejecting both a naturalistic and a

culturalistic sense of landscape (Macnaghten and Urry 1998). Architectural theory has also benefited from the concept of dwelling, as seen in Christian Norberg-Schulz's extensive discussion. He sees dwelling as structured by identification and orientation, with identification pertaining to meaningful relations between humans and things and orientation to how beings engage the space they constitute (Norberg-Schulz 1985). Also de Beistegui attends to the relation between dwelling and architecture when discussing the "interstitial spaces" found between modern calculable space and the lived, experienced place of dwelling (de Beistegui 2003, 139–68).

Part I

EXPANDING THE COMMUNITY

Chapter 1

Sameness

It is now time to explore ways in which the sameness of community or what is shared amongst political beings are potentially extended or envisioned as more inclusive of those formally not included. It was suggested in the introduction that the very issue of statelessness was symptomatic of several assumptions on the political community. Presuming a notion of sameness that presupposes difference as the condition of possibility of the community makes boundaries the primary object of politics. Pushing, expanding, redrawing, but also reinstating boundaries becomes what the meaning of politics revolves around. Boundaries secure that the unity of the community remains in place, so that the semblance of its members can be distinguished against the difference of non-members. While this forms a closure of the meaning of the political and of what it means to be a political subject, the primary concern of this chapter is how a community predicated on a notion of sameness that presupposes difference turns belonging in the political community into a question of membership.

It will be seen how also more inclusive accounts of the political community persist in upholding exclusions as long as they continue to presume a mutually implicating notion of sameness, difference, and identity as the condition of possibility of the political community. This chapter explores the implications the idea of membership has for how the political community persists in separating people, maintains inequalities at its core, contradicts foundational democratic principles, and endures in thinking of the community as organised on the basis of the differentiation between us and them.

THE PARADOX OF POLITICS

In our contemporary world, the ultimate formal proof of belonging in a political community rests in state citizenship. At a glance, aside from sharing the same citizenship status, there is nothing requiring the similarity of persons holding citizenship in a democratic state. However, this observation rests on the assumption that there exists no necessary unity of the people that is prior to the emergence of a state of which persons can be citizens. According to theories of democratic legitimacy, the boundaries of the political community must be democratically legitimised in order for the community to be democratic. This "paradox of politics", attributed to Rousseau, reveals how the assumed unity of the people and the challenges unification poses for political thinking are by no means a new conundrum (see further Connolly 1995; Honig 2009; Näsström 2007). In contrast to the other usual suspects of social contract theory, Rousseau refrained from placing power in the hands of a sovereign representative of the majority. Instead, Rousseau sought to mediate the will of all, emanating from the multitude and the general will of the people. While Hobbes and Locke were mainly concerned with the legitimacy of government, Rousseau turned his attention also to the predicament of the *legitimate foundation of the people*. How could men be both authors and subjects of laws, he asked when addressing the problem that "men would have to be prior to laws what they ought to become by means of them" (Rousseau, quoted in Näsström 2007, 630). If the people would decide on its own boundaries, it would have to "[assume] beforehand the very issue that is under consideration" (Näsström 2011, 116).

Since democracy refers to rule by a *demos*, by a people, it goes more or less without saying that the people constitutes the only legitimate source of political authority in democratic theory. However, although being the source of the legitimacy of authority, the people fails to legitimise itself. Hence, when the question of legitimacy is directed to the delineation of the people, the problem of it being both the source and the object of legitimacy arises. This forms a gap or a hiatus in democratic theory—in order to legitimise the mode by which it is circumscribed, the people would have to be prior to itself (Näsström 2007). Further, the procedures and requirements for inclusion set up by the *demos* will also be marked by that same legitimacy gap. These procedures are supposed to constitute the legitimate basis of the *demos* but can only derive their legitimacy from that very *demos* they are supposed to ground (ibid.). This problem forms a "continuous feature of modern political thought", Näsström claims (Näsström 2003, 820).

But if the people would have to be prior to itself, this begs the question of what grounds the boundaries of the *demos*. Arash Abizadeh has shown how this question makes democratic theory involuntarily fall back on cultural nationalism. For the latter, a shared culture would intuitively satisfy as a prepolitical ground of the people. Such a myth of a common culture is a myth of both unity and sameness as people are expected to share in the same cultures and values with the same form of appreciation and obedience. Yet such a shared culture can never fully exist. Even if vested in a common language, a shared culture will prove to be a hybrid of an array of influences, both when studied diachronically over time and synchronically in the present. A shared culture fails also to solve the membership question, that is, the problem of closure: "In practice, there will always be both internal variation in cultural features among members of the putative group and external overlap with putative non-members" (Abizadeh 2012, 870). Arguably, a shared culture can be fostered, but then culture turns into a political project, as something to be realised as one of the goals of politics, rather than providing for a prepolitical ground for political authority. Creating culture is in fact what cultural-nationalist projects always and explicitly are about, Abizadeh maintains. But if a shared culture is conjured through politics, rather than constituting its point of departure, something else must be sought to anchor the collective subject of politics. Here another myth of commonness steps in to provide for the foundation that culture fails to produce—the myth of ethnicity. This myth of common ancestry stands more forcefully against the changes of time and provides cultural nationalists with that prepolitical grounding their theory calls for. Cultural nationalism—which would demand the assimilation of any outsiders in order for them to be understood as parts of the common—thus collapses into a request for ethnic kinship when confronted with the questions of legitimacy and boundaries (ibid. 871–73). Hence, "behind the cultural nation lurks an ethnos, eager to cover its next of kin with a warm and rather constricting embrace. A constructing embrace" (ibid. 873).

Democratic theory is not free from this problem. The procedural character of democratic theory requires the legitimisation of the procedures governing the articulation of a collective will. Therefore, the unity of the people, and the procedures by which the collective will legitimately can be articulated, share the problem of being legitimised by the law—the social contract—which they should serve to legitimise (ibid. 874–75). Trapped in the circularity of legitimisation, traditional democratic theory turns to the assumed shared culture of cultural nationalism to find a ground for determining membership—"Incapable itself of legitimizing the prepolitically exclusionary boundaries it has traditionally presumed, modern democracy has relied parasitically on a cultural-nationalist principle of legitimacy, which equates the people with the cultural nation" (ibid. 875–76). And since the cultural nation cannot redeem democratic theory from its problem of legitimate delineation, it appears that democratic theory must also be on guard against the spectre of an *ethnos* hiding in the shadows of the *demos*. Citizenship, then, despite its formal character, becomes a question of membership not only in terms of who is in and who is out, but also in terms of who is sufficiently *alike* and who deviates too much from the nation sustaining the boundaries of the *demos*.

So why does this problem at all emerge? This book maintains that the circulation at work in the problem of a people that beforehand has to be what it ought to become by its laws rests in a prior assumption of unity as a necessary precondition for a viable political community. Presupposing unity as the condition of possibility of the people means that the circularity inherent to its legitimisation is bound to arise.

In want of a legitimised ground for the unified people, the nation has served a fruitful way to deepen the notion of the people when its boundaries fail being accounted for (Abizadeh 2012; Yack 2012, 5). With Emmanuel de Sieyès, the nation, as the source of both sovereign authority and of the *demos*, stepped in to redeem the democratic people of its legitimacy problem (Bartelson 2009a, 42). Thereby, the hiatus of democracy was not only circumvented, but, by placing the nation rather than the individual in the state of nature, Sieyès also concealed the legitimacy problem of the people and the constitution of its law (Näsström 2003, 817–18). Thus, the turn to history is consequent to a prior bracketing of the question of legitimacy of the unified people. Apart from installing involuntariness at the heart of democratic theory, alluding to history does not manage to make the problem of legitimacy go away, for how to decide on who gets the privilege to write history (Näsström 2007, 631–33)?

> Resignation to the contingencies of history does not at all fit with the rhetoric of popular sovereignty. Yet, in effect, that is what many liberal democratic theorists seem to demand from peoples uncomfortable with the shape of their communities; that they should accept whatever potential injustices history has served up to them with the boundaries of states so that we can all get on with the task of establishing liberal democratic forms of government. That this advice almost invariably comes from people who are quite comfortable and unexposed within the given boundaries of states, people who, in effect, are happy with the partners they were given when the music stopped playing at the dance of history, makes it harder to accept than it would otherwise be. (Yack, quoted in Näsström 2007, 633–34)

Instead of seeing history as a multitude of lines unfolding simultaneously—in separation, exchange, and intertwinement with one another—attempts to overcome the problem of a unified people by alluding to the nation or to history forward an unelaborated and unitary view of history. In this view, the present could be said to provide the point of departure for history. The present, or a particular view of the present, is what history is understood to inescapably have led up to. Stories or images that divert from or contest the particular interpretation of the present are excluded from a national history taken to support a circumscribed people. Such a history unfolds within both spatially and temporally restrictive frames. Spatially, the national history takes place within a particular delineation of territory that naturalises boundaries and safeguards them from external interference. Temporally, national history is framed and circumscribed by being expressed in terms of linearity (Stephens 2013). Hence, nationalist time posits a temporality that seeks to take control over its own political fate (Yack 2012, 135).

The linear understanding of history will always both be excluding and form a simplification by barring alternative or subversive stories of people-formation. Instead, the beginning is taken to relatively smoothly—or at least with a certain sense of direction—lead up to the present.[1] The circularity of Rousseau's paradox replicates also in the writing of history. To provide a unifying account of the people, its history is written with a linearity that wipes out all erring and going astray in the formation of the people. On the one hand, history is taken to be productive of a certain unity of the people; on the other, that unity forms the very source through which the history of the people is written.

The assumption of unity as a prerequisite for the community holds our understanding of historical unfolding in a tight grip. Comparing Germany and France, Rogers Brubaker argues citizenship in the nation-state to be "inevitably bound up with nationhood and national identity" and "membership of the state with membership of the nation" (Brubaker 1992, 182). In both cases, citizenship and nationhood emerge as inescapably intertwined; in the French case, the state produces a nation, whereas in the German case the nation results in a state. Brubaker concludes that citizenship is closely connected to nationhood, leaving membership in the community to be primarily a question of national belonging. However, the problem here is how the conception of the nation-state appears as "the only imaginable form of political organization, since within this ontological framework, history can only take place in the space between these two concepts, and historical dynamics are only comprehensible in terms of the permutations the concepts of the nation and the state allow in terms of their respective ontological primacy" (Behnke 1997, 245). Even if Brubaker strives to situate citizenship and nationhood in historically and culturally specific contexts, he still imposes the nation-state on his analyses as both the fundamental principle for, and teleological outcome of, political organisation. He thereby neglects how, to the eighteenth-century revolutionaries, the concept of the nation did not denote the cultural homogeneity and common identity it is marred by today but was foremost opposed to the privileges of the nobility and monarchy in the *ancien régime*. Hence, the nation was a "dynamic concept, with the French nation as the 'epicenter' of an ever-extending democratic, egalitarian, sovereign nation of free people" (Behnke 1997, 255; see also Douzinas 2007; Fitzpatrick 2004). By this de-historicisation of the concept of the nation, the nation becomes a "ghostly entity" that imposes itself on history and leaves few glimpses of alternative routes of historical unfolding (ibid. 257).

MEMBERSHIP IDENTITY

The assumption of unity as a precondition of the political community functions to make also expansionist notions of political community and belonging prone to exclusions. Classic cosmopolitanism, understood in terms of a world citizenship and a global ethics that acknowledges how both political policies and human relations exceed national boundaries, forms an obvious example of this. While on the one hand emphasising the possibility of simultaneous membership in both a national and global community and the possibility of cultural identity and group membership to flourish in concomitance with membership in a world community, cosmopolitanism also describes the need for universally shared ethical standards and for world government (Archibugi and Held 1995; Heater 1996, 2002; Held 2002).

However, attempts to identify a world citizenship that harbours local variation tend to neglect how understanding states as entities accommodating the cultural heterogeneity of mankind is deceptive, since the problem of closure emanating from the notion of distinct cultural units remains present. They further have to counter the risk of understanding culture in prepolitical terms. Since cultural change is unavoidable, especially when viewed over time, the recourse to an essentialist understanding of culture remains as an imminent peril. Similarly, assuming pre-existing universal values for the world community poses mankind itself as bounded by and to these values, something that inexorably bids for the externalisation of all those who cannot or are not prone to recognise an allegedly impartial account of a cosmopolitan moral law. Thus, prior to being a citizen of the world, one must first ascribe to the sameness conditioning a specific moral and institutional order. The difference sustaining such a notion of sameness would be found in all those communities who fail to live up to a predominantly liberal ideal of the political coexistence. Hence, the ability to claim oneself as a citizen of the world becomes in effect often dependent on the prior membership in a sufficiently civilised state.

The same kind of neglect for internal cultural variation both within the minority group and within the larger political community can be seen in multiculturalist approaches to citizenship. Affirming access to one's own cultural identity as necessary for the development of self-identity, communal belonging, and individual freedom implies addressing the cultural diversity within the national sphere and recognising particular minority rights for particular groups (Kymlicka 1995; Parekh 2006; Young 1989). If the latter is the multicultural aim, the issues of how to decide on group boundaries and how to engage difference internal to groups become central. A solution would be to settle for culture as a relatively homogeneous and stable identity, whereas differences in personal identity, such as gender, sexuality, or religion cut across various group identities (see, for example, Kymlicka 2001).[2] This, however, neglects the cultural variations in, for example, gendered experiences. It also disregards the cultural diversity stemming from experiences of life in the diaspora.[3]

Both above approaches draw citizenship as a form of membership based in a particular national or cultural identity. Inclusion in the political community will either rely on the belonging to an assumed pre-existing sameness of cultural or national identity or be fostered through assimilation into a common ethics. Walker's observation of how "most contemporary discussions of multiculturalism and difference already assume a singular culture, or community, in which cultures and differences can be understood" (Walker 1999, 185) is therefore equally applicable to cosmopolitanism, the main difference being the scale on which it operates.

The discussion on *ipse*- and *idem*-identity helps grasp what is at stake in understanding group-belonging as preconditioned on sameness, unity, and identity. Identity understood as selfhood, as *ipse*-identity, should function to ground self-reflecting political subjects who can reflect on their own actions and are in charge of their own decisions. Yet, with respect to a collective, such a notion of selfhood serves also to ground a *collective self*. As a collective self, the community is vested with an agency that transforms the issue of inclusion into an issue of self-inclusion. Thus, the collective can initiate a discourse on 'us' and what is meant by 'us' and 'ourselves'. Here begins a discussion on who can be characterised as belonging to 'us' and by what measures a being can be understood as belonging to 'us'.

With the unelaborated and relatively homogeneous understanding of culture presented above, this understanding of identity as *ipse*, or selfhood, becomes merged with *idem*-identity. Lindahl understands collective *idem* as a closure, as a question of grasping what is same in relation to itself and different with respect to other collectives. There are two consequences to this. First, this deviates from the identity as subjected to the possibility or ability to think of oneself and self-reflect, since self-reflection presupposes precisely that one is *not* completely the same in relation to oneself. The collective thereby emerges as relatively homogeneous. Second, describing the character of the collective, the merging of *idem*-identity with *ipse*-identity means that the selfhood of the collective now surfaces as requiring particular qualitative features of its members. With Ricoeur, it can thereby be seen how the question of who can belong to the community becomes transformed into a question of what particular qualities can be acceptable in order for one to afford membership in the community.

Now, understanding the issues of integration and assimilation on the basis of *idem*- and *ipse*-identity suggests that integration describes the issue of how to make possible inclusion of beings in the community on the basis of their selfhood. It is a question grasping the people as a plurality able to accommodate the selfhood of each individual and grant that their *ipse*-identity can be sustained. Assimilation, on the other hand, implies the requirement for any putative member of the community to comply to, or assimilate to, the particular characteristics of the *idem*-identity of the collective. Thus, putative members of the political community would have to be able to conform to what is considered as the sameness of the community while at the same time distancing themselves from what is considered different or other to that community.

Cultural nationalists would adopt assimilation as a route to membership in the political community, while also tackling the problem described by Abizadeh of how to delineate the cultural nation without taking recourse to an idea of *ethnos*. With their emphasis on liberty and equality, liberal democrats, on the other hand, strive for integration as the form through which persons become perceived as parts of the political community. Yet, as they continuously allow *idem*-identity to operate alongside *ipse*-identity in their theories of citizenship, they become prone to reinstall sameness as a precondition for membership. Thereby, despite their ambitions, they describe the process of becoming part of the political community as a process of assimilation rather than integration.

To present a liberal argument for multiculturalism, Kymlicka sees access to one's cultural identity as necessary for the development of self-identity, communal belonging, and individual freedom. However, when turning from the rights of national minorities to the situation of immigrants, the latter are presumed to have voluntarily uprooted themselves from their own culture and thereby to have left any aspirations or claims to that culture behind. Therefore, "the good of cultural membership for immigrants is primarily a matter of enabling integration" (Kymlicka 1995, 114). This effectively changes the right to one's *own* culture to the right to *a* culture. It also diverts the discussion from an endorsement of the importance of selfhood as a crucial aspect of human existence, to a revalorisation of *idem*-identity or the sameness of a liberal political identity. By consequence, cultural membership for immigrants becomes equalised with assimilation into existing categories. Abizadeh's reasoning can here be rephrased; to assume that unity forms the binding condition of possibility for the political community allows the ideal of selfhood to be merged with the sameness of *idem*-identity. But as *idem*-identity operates according to resemblance and by delineating against difference, the question of the political community becomes an issue of determining particular qualitative features of its members. Presented as an issue of membership, to which standards one needs to comply, citizenship will thereby be haunted by the spectre of national character as grounds for belonging.

The slippage between integration and assimilation can be discerned also in perceptions of the workings of identity as selfhood. Using the French ban on headscarves in schools as an example, Seyla Benhabib notices that the adoption of cosmopolitan norms allows young Muslim women to oppose a state that circumscribes their religious freedom. Allowing the women to retain their traditional identities, cosmopolitan norms function simultaneously as a possibility to empower the young women to renounce their assumed designation to the private sphere. Through their enactment of cosmopolitan ideals, the women can both "talk back to the state" and transcend their traditional positions and are thus "transformed from 'docile bodies' to 'public selves'" (Benhabib 2004, 209).

With this, Benhabib narrates a story of a universal cosmopolitan moral, which excludes possibilities of political struggles that do not employ cosmopolitan norms (Honig 2006, 113). To Benhabib, the struggle for equality and freedom of religious expression becomes vested in the sameness of a cosmopolitan moral, as becomes the very selfhood of young Muslim women. Cosmopolitanism functions as the possibility *par excellence* to express dissent and "talk back" to a discriminating state. Indeed, the girls need to transcend the *idem*-identity expressed through their religion in order to not be merely "docile bodies". "Muslim women" appears as a uniform category that allows for little variation and which offers few grounds for political engagement. Instead, it is a cosmopolitan moral that grants them the selfhood that lets them function as acting subjects.[4] However, exactly through locating *ipse*-identity in a cosmopolitan moral and denying a similar understanding of selfhood to be vested in the identity of the young Muslim woman, Benhabib will inscribe cosmopolitanism as also an *idem*-identity. Also cosmopolitanism will therefore emerge as an identity describing particular qualitative features of its bearers, which, in this case, takes the form of the very feature of selfhood.

CITIZENSHIP IDENTITY

Even in cases when identity or collective identity are elaborated as diverse and multi-layered, the idea of sameness and unity resurfaces through a continued affirmation of a 'we' that grounds the community and provides it with selfhood.

Notions of multiculturalism see identity as malleable, fluid, and multidimensional and as requiring difference-based policy responses that are sensitive to such diversity (Modood 2007). To meet a kaleidoscopic image of society, in which one experiences multiple group belongings that reflect a diverse array of aspects of one's person, multiculturalism must accommodate a form of integration that opens for the contact, experimentation, and transformation of cultural belonging (Modood 2007; Parekh 2006). For these authors, the question of political community unfolds around the issue of how to create a community that is open, flexible, and indeterminate, while still grounding the unity of the people. To manage its own plurality, the pull of a common national identity is thought to provide the unity allegedly needed for the community not to fall apart. With this, the perspective that identity should be respected simply because it is of importance to its bearer is turned into an understanding of identity as central to an individual's choices and commitments and that shared identity grounds moral commitment.

Similarly, the understanding of embedded cosmopolitanism embraces the indeterminacy of identity and centres on how the moral agent must be understood as radically situated. The source of morality lies neither in a common humanity nor a bounded community. The intersection of various identities is instead what reveals the indeterminacy of identity and allows for a moral concern that extends beyond one determinate category. Yet the reason why the moral agent is capable of showing ethical concern beyond her own primary sphere is not because of a *general* experience of identity as multifarious and indeterminate, an experience which in itself would expose any other individual as constituted by an equal complexity. Rather, it is the identification of particular identities common to the moral agent and its other that paves the ground for an outward-reaching moral (Erskine 2000, 588; 2002, 477). Hence, my moral concern can only reach as far as I can recognise something of myself in the other. If the other is too different from me, if she has no identity intersecting with my own, she is beyond my moral scope. Drawing on intersecting and overlapping identities means to bar what is incalculable to a particular realm of ethics from the reach of moral concern. Once again the spectre of a notion of sameness that presupposes difference emerges as an underlying presupposition for political coexistence, as does the formally already included as the primary moral and political agent.

Again, a notion of sameness that presupposes difference emerges as the condition of possibility of the political community. Rather than the shared struggle to make a viable democracy function in the face of plurality, or the realisation of the multi-layered existence shared amongst beings, is it the sharing in the same overarching national identity or the same identity ascriptions that grounds moral commitment or the possibility of a functioning political community.

Hence, the question of political community becomes premised on identifying a particular representation of 'us' in order for the community to be operational. A continued critique against multiculturalism is that the demand for equal recognition and minority rights will assure the persistence of categories established through discriminatory structures. What also persists when making subsumption under a category necessary for rights adjudication is the assumption of groups as bounded by a notion of sameness that presupposes difference.

Consequently, even if Iris Marion Young's refusal to resort to *any* assumption of an overarching identity makes her take on multiculturalism more radical than the above, her insistence on minority rights makes her vulnerable to the same critique as levelled against liberal citizenship. Founding a notion of citizenship in "what citizens have in common as opposed to how they differ" is perilous, Young argues, since a common good yields to a homogeneous citizenry and can never reflect a differentiated society (Young 1989, 250). A universal citizenship will in fact allow "equality" to be "conceived as sameness" when disregarding the differences in status, power, rights, and wealth within civil society (ibid.). Correspondingly, a law that sees everyone as the same and applies accordingly is blind to individual and group differences. The idea that citizenship is the same for all has thus "translated in practice to the requirement that all citizens [should] be the same" (ibid. 250–54, at 254).

Notwithstanding, Young has to wrestle with the issues of, on the one hand, accounting for a heterogeneous public through opening up spaces for their voices to be articulated, their claims to be raised, and their experiences to be circulated. On the other, she is bound to resort to the tools given through democratic politics and rights adjudication. Formulating an institutional response to inequality makes Young rely on recognition under particular categories as a condition of possibility for equality. As maintained by Lindahl, legal collectives require both *ipse*- and *idem*-identity since they need to determine the self to which a right or law applies and ascertain that a right or law applies in the same way to each one in the same collective (Lindahl 2014, 91–92). Therefore, despite emphasising the flexibility of identity, Young reaffirms both *ispe*- and *idem*-identity when drawing her notion of differential justice. Further, she must also resort to a conceptual thinking that seeks to overcome former injustices by using the same categories that were crucial for them to emerge in the first place. While Young's identity politics might present a welcome route out of a society where privilege is given to a liberal citizen subject, it still must strive towards the upheaval of its very basis—identity and group belonging—in order to avoid membership, exclusions, and a notion of sameness that presupposes difference as preconditions of the political community.

Their differences aside, both Young and Benhabib rely on a linear and progressive temporality, which assumes that rights adjudication, universal cosmopolitan categories, or minority rights, as well as rights-claiming *per se*, will over time solve the difficulties of securing the rights and status of non-included. Through an evolutionary understanding of time, Benhabib can explain any shortcomings of human rights as due to an imperfect but yet progressive cosmopolitanism. This neglects those not encompassed in cosmopolitan claims, which form remainders "produced by the conditional order of universal hospitality" (ibid. 113). Multiculturalists, on the other hand, are highly aware of the remainders and their source, but when revalorising formerly excluded categories, they also maintain these as part of the political landscape. Thus, the question of the political community is intensified as a question of determining boundaries of inclusion and exclusion. Rights adjudication dependent on subsumption under a category also needs to wrestle with the problem of remainders generated by conceptual and representational thinking.

The reliance on identity, even if elaborated as fluid and malleable, makes multiculturalists and cosmopolitans alike vulnerable to a heightened sensitivity to the radically Other, described by Balibar and Prozorov. A politics relying on identity ascription will intensify the principles of proximity and distance between cultural experiences. It also risks presuming a civilizational difference between what is compatible and incompatible. Indeed, the very practice of citizenship could be perceived as a disciplining technique that marks out a civilizational difference between the reconcilable and irreconcilable. Relying on the partitioning of the global population into discrete units, citizenship conditions the forms for political organisation and produces international hierarchies (Hindess 2000). When this highly productive management of population is coupled with a teleological account of citizenship, it both conforms political existence to one ideal and sets populations against each other.

While the economic and social security attributed to citizenship is something that pertains principally to citizens of prosperous states, the vast majority of people are citizens in poor states, meaning that the benefits assigned to citizenship are greatly misleading (ibid.). Still, being a citizen of some state is generally better than having no citizenship at all. Coupled with a teleological discourse, by which citizens and their states develop on a scale that will make also the poor progress towards the prosperous, citizenship "provides states with an internationally acceptable rationale for regulating the movements of those who appear (or threaten to appear) on or within their borders as refugees from war and other forms of institutionalized violence or simply in search of what they believe will be a better life" (ibid. 1496). Hence, the one's not yet touched by the blessings of the modern democratic state can be kept with the hope of doing so in time; "if only they would stay at home and learn to behave themselves, they too could be citizens like us" (ibid.). The progressive time of citizenship now teams up with a spatiality of the inside, by which populations and development are confined to particular places. Here, the very category of the citizen becomes burdened with differentiations; the idea of citizenship becomes internally divided to the extent that public discourse may deem citizenship in some states irreconcilable with the ideals of citizenship in others.[5] Thus, citizenship *itself* emerges as an *idem*-identity, not only providing the individual with a sufficient degree of liberty and equality to be in charge of her own life and make decisions as a free self, but to the extent that citizenship is vested with a number of qualitative features that structures an inside and outside of international society.

THE PARADOX OF DEMOCRACY

Lindahl underscores how a collective self engages in self-inclusion, that is, in establishing an idea of the selfhood of the collective and operating so that this selfhood is sustained also through practices of inclusion of new members. When Benhabib understands cosmopolitanism as a project of mediations between universal norms and local particularities, she makes an important effort to attach a flexibility to both the community and its values and members (Benhabib 2004, 2006). Still, she affirms the *demos* as the chief political actor, when concluding that it is only the *demos* that can "change its self-definition by altering the criteria for admission to citizenship", as "the line separating citizens and foreigners can be renegotiated [only] by the citizens themselves" (Benhabib 2004, 206). Whereas she emphasises how challenges and re-articulations of rights often emanate from new groups of rights-claimers, the latter are in fact denied the agency to put the reach of liberal rights to test.

This, however, assumes the *demos* as bounded. Benhabib affirms that a clear understanding of who are to be counted as included in the *demos* is a necessary feature of a functioning democracy; "the logic of democratic representation … requires closure for the sake of maintaining democratic legitimacy" (ibid. 220). However, the all-subjective principle in democratic theory, which states that anyone subjected to political power should have the possibility to be part of the deliberative process by which that same power is established, suggests that the boundary of the people *cannot* be a question only for the already included citizens of the community (Abizadeh 2008). On the contrary, decisions taken regarding the boundaries of the community would be legitimate only if justified by members and non-members alike, since they affect also non-members. By consequence, any regime of unilateral border control suffers from a legitimacy gap from the point of view of democracy. Taking this problem seriously means to acknowledge that the people includes each and every individual subjected to the political power exercised by the people, and that this, at the boundary, would encompass members as well as non-members of the political community. Thus, the all-subjective principle would grant everyone subjected to a decision on the boundary of the people a right to have a say in that same decision (ibid.).

As indicated by the quotes from Foucault in the introduction, identity as selfhood can, when applied to a collective, harbour traits of the sameness of *idem*-identity. For Abizadeh, this becomes apparent when he sides with Habermas to find an articulation of the procedures, processes, and regulative standards of democratic rule, while simultaneously noticing how "no actual set of procedures and processes can fully live up to the constitutive ideals of democratic self-rule. ... The specific procedures in place *will significantly impact political outcomes*" (ibid. 880, my emphasis). Here, the inability to completely pin down a representation of the people comes as a saving ground. With its shifting character, the people would serve as its own regulative ideal; the people is "bounded only by the capacity of communicative decision-making practices" (ibid. 881). Nevertheless, this does not solve the problem of circularity since the *capacity itself* suffers from the same legitimacy gap, as does the *demos*. However understood, the capacity to make decisions will always *in advance* have delimited possible formations of the people. The notion of selfhood, of *ipse*-identity, implies a subject capable of conducting and orienting itself in a particular way with a certain degree of self-reflection and insight. It also implies that a subject is sufficiently free to do so. To not further scrutinise the gatekeepers of democratic processes indicates the neglect of the potential externalisations and remainders to any assumed universal account of political existence. However thin in character such a universalism is, there will always exist diverging accounts of what it means to engage in political life and how to understand political belonging and political community. Thus, also in attempts to describe the political community solely in terms of *ipse*-identity, the contours of a constricting—and also rather constructing—unification remain in place.

A different path would be to explicitly forward democracy as a form of rule that "holds an absence of unity at its heart" (Näsström 2010, 213). This withdraws from the many attempts to use the *demos* as it transpires from its national character as a blueprint for understanding a global *demos*. Instead, an absence of unity of the people forms the very "condition of possibility for a democratic society" (ibid. 212) and forwards the *demos* as a forum for disagreement. Absence of unity is also what distinguishes democracy from other forms of rule, as it embraces "the plurality that goes with an authority that refuses to become harmonized into a single voice" (ibid. 213). Plurality becomes thereby what democracy is all about, since the "numerous" and the "countless" is what democracy is required to equally account for to "give the excluded and silenced a political voice" (ibid. 207).

However, while addressing the *incalculability* of the people, Näsström does not reject *counting* as a necessary element of democracy. To the contrary, the lacking unity of a people with a multitude of opinions is what on the one hand safeguards equality but on the other makes it necessary to equally count the numerous voices of this incalculable authority. With this, the problem of finding a measure for whom to count and for how to carry out this count arises. Should it be "fellow citizens, Europeans or human beings" that count as parts of the *demos* (ibid. 214)? To Näsström, this opens the gates for the debate on the people, meaning that the enabling condition for democracy is upheld. Still, this means that she also lets the people remain within a rationale of calculability by placing the incalculable within reach of calculation. That is, she keeps the problem of who gets to be counted and according to what measure firmly in place, by letting this problem be a condition of possibility for democracy. However, this means that sameness expressed in the form of membership also is kept in place.

Thus, the circularity Näsström sees operating in the legitimation of the community is reproduced in the relation between the incalculable people and a democracy that counts. The condition of possibility for the democratic people is the incalculability of the people; it is the multitude of opinions and wills that forms the democratic people and that lets the will of the people emanate as the will of the *demos*. However, the condition of possibility for *deciding* on the will of the people is calculability, that is, that which forms the condition of impossibility for the democratic people. Hence, even if Näsström evades the unity of the people, she maintains a unifying authority as a built-in feature of democracy. Therefore, a notion of sameness that presupposes difference is again assumed as necessary for a viable democratic community, leaving the paradox of the legitimacy of a unified *demos* to emerge in the new costume of the calculable incalculability of the democratic people.

Despite their attempts to move away from state-centred notions of community and from unilateral border control, both examples above reveal how democratic theory hinges on a state rationality of calculability. This is due to how democratic theorists rely on equality as inclusion. The problem is that equality, whether in number or worth, rests on an assumption of sameness that presupposes difference. Equality of number counts each subject once but has to ascertain that a subject indeed is countable as a subject. Equality as worth seeks a measure for what counts as *logos* but must then differentiate against what is not deemed consistent with the conceptions of rationality. Inviting a notion of sameness that presupposes difference through the back door, both notions of equality leave demo*cracy* to eventually undo the possibility of an incalculable *demos*. Thus, since democratic theorists in one way or another confess to equality, they are bound to give way to distinct communities and turn the problem of inclusion into a question of membership.

CONCLUSION

This chapter has seen how the assumption of unity as prerequisite for the political community has conditioned political existence and agency on prior membership. Despite their attempts to broaden the scope for inclusion, theories of citizenship above have maintained strict delineation between insiders and outsiders. In the above we have seen different attempts to incorporate a multi-layered account of identity in the thinking of political community. Yet, these attempts have in one way or another referred back to *idem*-identity, to identity as sameness determined on the basis of particular qualitative features. Even *ipse*-identity, it was argued, was merged into *idem*-identity when seen as the crucial source for claiming rights and liberties. As a consequence, political belonging becomes understood as membership.

The introduction dwelled upon Arendt's observation of how groups of stateless persons were symptomatic of our contemporary political situation. It was suggested that statelessness, understood in terms of how formally non-included beings lack political and legal standing in the places where they reside, was symptomatic of a political imaginary that understands political belonging as a question of membership. For such an imaginary, the question of political community is a question of who and according to what parameters. As seen above, assuming unity as the condition of possibility of the political community, both multicultural and cosmopolitan citizenship theories are bound to give way to new exclusions despite their attempts to open up the meaning of community. Embracing the idea of identity as a category not only crucial to one's personal life, but also as central for political action and inclusion, means that these theories will retain a distinction between inside and outside. Thus, separation between peoples persists, both in terms of between different political communities and in terms of those who have and those who lack a place to call their own.

Vested in the notion of identity, membership in the community will be grounded in an assumption of sameness that presupposes and installs differences. A thinking of the community as organised on the basis of a differentiation between us and them thereby endures. So does inequality as a central feature of the political community. The problem of inequality emerges in two forms. On the one hand, it consists in the inequality between formally included and formally non-included, where the latter can take any part in the community only through the benevolence of the former. On the other hand, the very process of being granted membership is vested in inequality, since assuming identity as a ground for inclusion turns integration into assimilation. Rather than forming a project, or even struggle, that moves the position of everyone taking part in the political community, integration is turned into the requirement for newcomers to assimilate into the existing order. Certainly, there is a great variation in the degree to which this is the case amongst above-studied theorists. Yet allowing a notion of sameness that presupposes difference to sustain the idea of the political community means that even more nuanced accounts of political identity will be haunted by demands for assimilation.

As a consequence, citizenship as a practice of inclusion will fail to live up to the foundational democratic principles it rests on. If democracy relies on the idea of rule by the numerous, then the *incalculability* of the people needs to be maintained to guarantee an abundance of opinions and perspectives. Adopting assimilation, rather than integration, constitutes a move away from such incalculability of a multitude of people. Assuming that democratic authority relies on the premise that each voice counts equally, democratic theorists indeed employ their own measure that will answer the question of not only who but what constitutes the political community. This measure is the measure of equality. Confessing to equality, whether in worth or number, means that democratic theories that draw their very strength from the indeterminacy of the *demos* are bound to give way to distinct and delineated communities. Therefore, despite seeking to avoid boundaries of the people, their entire understanding of politics becomes structured exactly around boundaries. It is to this problem the next chapter turns.

NOTES

1. Constantin Fasolt further addresses the overall problem of writing history. History can never escape being viewed through the looking glass of the present; the present constitutes the prism by which any knowledge of history must be read. Therefore, the motives, the drives, and the wills of the present will mark—and mask—the past that we seek knowledge of (Fasolt 2004).

2. Feminist citizenship theorists point out how an understanding of cultural groups as relatively homogeneous risks overlooking injustices against women within minority groups (Saharso 2003, 200; Yuval-Davis 1997a, 77). They further address the resemblance between minority identities and national identities in their construction of women as cultural symbols of the collective (Yuval-Davis 1997b, 67).

3. A common feminist and liberal approach to the standing of women in various minority groups endorses gender rights as cutting across different groups and as countering potential patriarchal tendencies within them. However, such feminist critique allows *one* image of the women to encompass the variety of problems experienced among different groups. For a discussion on the problems with such an approach with respect to the discrepancies of experiences of women in the West and in postcolonial cultures, see Chandra Mohanty (Mohanty 1984).

Rather than assuming that women experience both the same form of subjection and the same form of emancipation, a gender-sensitive theory of multicultural citizenship would have to address the difference between women in different groups. Indeed, Linda Zerilli shows how a freedom-oriented feminism must refrain from any subject-centred claims or determinate understanding of the concept of the woman (Zerilli 2005)

4. Honig critiques Benhabib's theory for being circumscribed by a linear account of temporality that makes that category of cosmopolitan agency, as well as of Muslim women, be lodged in homogeneity. Further, cosmopolitanism escapes being assessed as an idea with ambiguous implications. For example, its role in the demand for minority assimilation in France in the first place is left uninvestigated.

5. A testimony to this could, for example, be found in the intensified rejection and denunciation of unaccompanied refugee children, predominantly teenage boys of Afghan descent, in Swedish political debate.

Chapter 2

Logos

The previous chapter suggested that by confessing to equality, democratic theorists give way to a notion of the politics as structured around boundaries. Thus, seeking to address the paradox of politics, that is, the paradox of a legitimate *demos* that should be both the source and author of the laws stipulating its constitution, democratic theorists on the one hand ascertain equality as a fundamental aspect of democratic legitimacy. On the other hand, precisely by doing so, they also install inequality at the very heart of democracy. Since a measure for equality will never be conclusive, remainders will persist. A possible conclusion of this would be to suggest that statelessness as a symptom of contemporary politics is symptomatic of an obsession with political equality. However, what would prove more correct is to say that statelessness is symptomatic of the adoption of an understanding of equality that is conflated with a notion of sameness that presupposes a notion of difference, which ultimately can be referred back to identity. As a consequence, the meaning of politics will revolve around the issue of boundaries.

The introduction suggested that statelessness as a symptom was symptomatic of how the meaning of politics is structured around boundaries. This chapter explores how boundaries become what politics is all about. Whether bridging, crossing, overcoming, or temporarily erasing them, boundaries persist as the very rationale of politics. This suggests that with equality comes inequality, with inclusion comes exclusion, and with sameness comes difference. Loosening the hold of established membership as a fundament for belonging in the political community, the meaning of politics will still delimit against a distinct outside, due to its reliance on transformative political acts that tie into the *logos* of political community. Hence, whereas in the previous chapter the question of political community revolved around the subject of *who* and *with what characteristics*, here, instead, the question of political community will revolve around the issues of *by what capacity* and *by what means* beings will emerge as political beings.

PRODUCTIVE PARADOXES

An approach by agonist political theorists to the problem of the legitimacy of the democratic people has been to embrace the paradox of politics and use it as a means to keep the meaning of the people open to dispute and change. Honig characterises this as "that irresolvable and productive paradox in which a future is claimed on behalf of peoples and rights that are not yet and may never be" (Honig 2009, 130). For her, as well as for William Connolly, the indeterminacy of the people forms the impetus for continued renegotiation and transformation of the meaning of the people and of political engagement.

Connolly encourages an ambiguous orientation to identity, aware both of its fluid, transmutable character but also of the restrictions and violence inherent to identity. He notices how the relation between identity and difference is such that the differences not encompassed by a particular hegemonic identity are taken to threaten the integrity and certainty of that identity. Differences therefore affirms identity, meaning that political tactics aimed at securing a hegemonic identity involve constituting hegemonic identity against the differences it does not include. As a consequence, a notion of difference turned into strict otherness becomes, on the one hand, completely externalised from the hegemonic common identity but forms at the same time a central support for the latter. Hence, although a unified community and a common identity is supposed to protect against fragmentation, violence, and anarchy, Connolly finds the latter problems to be exactly the product of identity regimes (Connolly 1995, xxi).

Despite being contingent, collective identity presents itself as harmonious and cohesive. A state campaigning against any elements disturbing to a common identity plays a crucial role in this. It turns difference into a "welfare class", marked by its disposability. This lets the appearance of democratic accountability be upheld, as the electorate can replace previous failures vis-à-vis the disposable welfare class in question. However, if the electorate should continue to have a choice concerning whatever threatens their common identity, it is crucial that difference in the form of an identifiable object upon which a politics of disposability can be enacted is maintained. No matter its composition, the welfare class thus becomes a "dispensable subject of political representation and an indispensable object of political disposability" that serves as a demonstration of power and political efficacy (ibid. 208).

For the neoliberal politics of the contemporary Western state, this welfare class has found a domestic expression in people living on social benefits, the unemployed, or anyone else drawn as a freeloader or outsider. The current turn to a presumed *external* welfare class through the commonly called 'European refugee crisis' has turned this social fabric inside out. While previous domestic margins have been allowed into a nationalist centre, a disposable welfare class has been made up of asylum seekers or poor migrants within the European Union. The very use of the possessive 'European' indicates the support an external group forms for the consolidation of what is made to appear as a harmonious, cohesive identity and also for the identity of a decisive state that can take firm action in order to uphold democratic accountability in the eyes of its electorate. This, however, obscures the aspects and failures of the state that are productive of any such class to begin with. It also obscures how a required construction of otherness easily can be transposed from an external to an internal group.

With an assumed homogeneous identity, deviant but of the same class of uniformity as those of the members of the welfare state, refugees are dehumanised and turned into flows, floods, or volumes, measured against what calculated abilities the welfare state has in order to maintain their existence. Thus, the welfare state can simultaneously present itself as the legitimate administrator of the justice, equality, and freedom of those it claims as its proper subjects *and* as concerned with the well-being of others by forcing them to stay put (elsewhere), where they too might experience at least some of the freedoms of the West. Not only, then, does citizenship create a form of global governance and ordering of national space but also electoral politics that entail the risk of becoming "a closed circuit for the dogmatism of identity through the translation of difference into threat and threat into energy of the dogmatization of difference" (ibid. 210).

For both Connolly and Honig, democracy itself forms the solution to hegemonic collective identity. Democracy inserts dissonance into the unity of identity and interrupts its dogmatisation. A democracy oriented to the problems of its own closures should be engaged by a critical responsiveness open to new movements and affirmations before they have turned into the entrenched identities that render them credible (Connolly 1991, xxix). Putting to use an ethos of pluralisation, deconstructing what is perceived as certain, and exposing it to questioning and re-articulation, democracy inserts dissonance into the unity of identity and interrupts its dogmatisation (Connolly 1995). Emphasising a reorientation of the ethos and energies of democratic politics, the bounded *demos* remains as the impetus of democracy for Connolly. Limits form the very condition of possibility of pluralisation and are that which sparks political action. Here, the meaning of politics and the meaning of the people coincides. Even if the ethos of pluralisation is not reducible to a fixed code and does not distinguish any clear end, its embrace of contestation and renegotiation demands at least temporarily settled positions.

Boundaries are that which animate politics, but they are also what animate the political community.

Both identity and difference remain necessary for the political constellation we inhabit, as well as for our ability to affect this constellation. This forms both a promise and a curse. Walker addresses this problem when underlining how "modern social and political theory has become intimately concerned with … a specific form of an opposition between identity and difference" (Walker 1993, 114). In the "plural world of change", the "truth of the one is opposed by the illusions of difference, whether of the many or of the realm of becoming" (ibid. 115). Such *illusions* of difference concern not the extent to which there are things different from the one, but rather that the different should follow another *structure of truth* than does the one.

Connolly's politics of becoming is a politics of the world; it emanates from the world and is directed towards this world. It engages the truths of the world, unmasks their contingencies, and diverts their effects. But this world we encounter is but one of possible ways of organising coexistence. The aim of Connolly, who also admits to this insight, is not to picture it as the best of possible worlds, but rather to *make the best of it*. Dedicated to doing so, he invites the non-included to the realm of political action and draws attention to the violence sustained by boundaries. For all his awareness of its problems, Connolly offers no escape from a notion of identity that presupposes difference and vice versa. What he presents are rather ways to engage productively in a politics that is attentive to the exclusions emanating from thinking the community as predicated on a notion of sameness that presupposes difference—a politics structured around boundaries.

Drawing on James Madison, Honig sees democracy as "a form of politics in which power is not received by grateful subjects but rather is taken, redistributed, reenacted, and rearticulated by way of liberty, that is, by way of poplar political action" (Honig 2001, 99). The paradoxical character of democracy as a form of rule where we are "living with strangers under a law that is therefore alien" and in which we are "being mobilized into action periodically with and on behalf of people who are surely opaque to us and often unknown to us" is precisely that which fuels democracy and democratic engagement (ibid. 39). To not realise this, but rather insist in the unity of the nation and on an 'other' to become a part of an 'us', would mean to promote an unexpanded and untested notion of democracy, which results in "the loss of opportunities to pass one of democracy's strictest test, the challenge to work and live and share not just with people with whom we have a great deal in common but also with those with whom we just happen to be bound up" (ibid. 117–18).

Rearticulating the question "How should we solve the problem of foreignness?" as "What problems does foreignness solve for us?", Honig seeks to divert apprehensions of foreignness as a necessary evil that politics is stuck with (Honig 2001, 4). Noticing how political theory has framed the foreigner as founder, as immigrant, and as citizen, Honig sees the foreigner in each instance resurface as both a giver and a taker; as both a supplement and a threat to national community. As a giver the foreigner provides something. The foreign-founder allows for the consolidation of law in the face of a general will that never can animate itself. The foreign-immigrant sustains the identity of the community in which she settles, either through the acceptance and support of the community suggested by her assimilation or through the consolidation of the communal boundaries that are confirmed by her refusal to integrate. Both as a 'good' and a 'bad' immigrant, she renders the boundaries of the community firm and sustains the identity of its members. In the image of "supercitizen immigrant", finally, the hardworking foreigner who is equally engaged in her family and her community is turned into the adored bearer of the narrative of an immigrant America, when she lends her flesh and actions to the story of the land of possibilities (ibid. 77–78). The immigrant as a giver to the nation sustains also the image of the immigrant as a taker; "the xenophilic insistence that immigrants are givers to the nation itself feeds the xenophobic anxiety that they might really be takers from it" (ibid. 99).

Attempting to avoid the above images of the foreigner, Honig redirects the question of the political community into thinking about immigrant relations to democracy rather than the nation. In this, she finds "taking" to be "the very thing that immigrants have to give us" (ibid.). If democracy is a form of politics in which power is taken, redistributed, and reenacted by way of liberty, then the foreigner, understood as a taker in terms of 'taking on' political agency, is moved into the centre of democratic politics. When "taking rights and privileges rather than waiting for them to be granted by a sovereign power", the foreigner turns citizenship into a *practice* rather than a juridical status (ibid.). Rather than based on membership, Honig draws a political community constituted by world-building and frame-shifting acts. Such acts denationalise the state and find instead "alternative sites of affect and identity" (ibid. 105). The one site of attachment that the nation-state poses is therefore transferred into several sites of empowerment and attachment (ibid. 122).

What Honig proposes in *Democracy and the Foreigner* is not simply that we can learn and enhance our democracy from the encounter with the other. Rather, she points at the encounters with the other as what *democracy is all about*. When Honig diverts the issue of the foreigner's relation to the nation to the question of the foreigner's relation to democracy, she also redirects the question of political community to a question of agency and action rather than of identity. Such a move is highly productive and commendable and plays on the same register as Engin Isin's notion of activist citizenship, which highlights how citizenship cannot be assumed simply as membership in a state. To Isin, more than being an insider, being a citizen means the mastery of "modes and forms of conduct that are appropriate to being an insider" (Isin 2009, 372). In contrast, the activist citizen does not simply follow script but creates her own scene that will redraw the boundaries between "citizens (insiders), subjects (strangers, outsiders) and abjects (aliens) and how these actors are to govern themselves and each other in a given body politics" (ibid. 384). For example, enacting themselves as having the rights to claim rights, *sans-papiers* do not foremost make a claim to public participation. Instead they expose the contingency of political categories and make actors emerge not as "already defined" but through modifying ways of being with others (ibid. 380–83, at 382).

However, both for Honig and Isin, identity, sameness, and difference remain in the shadows of their attempts to redirect the question of political community from one of how to determine membership to the issue of claiming belonging through action. Peter Nyers raises concerns about how Honig excludes what he calls the abject foreigner, like the deportee, the failed asylum applicant, or the overstayer (Nyers 2003, 1079). Taking the sentiments of the already present community as her point of departure, it seems that Honig omits those who, rather than being not yet included, are already excluded.

Likewise, Isin relies on the *public* character of citizenship. Even if activist citizenship is taken to disrupt historical sites of citizenship such as masculinity, warriorship, and property, there is a certain purposiveness inherent to acts. Acts are public and decisive; they create a scene. However, being purposive, decisive, and public rather than private, vulnerable, and precarious, acts of citizenship seem to align more with the historical masculine images of the citizen warrior than Isin most likely would wish.

Further, activist citizens put to work an incalculability that destabilises given boundaries and recasts categories of the community. Activist citizens indeed engage in the casting of given categories, as their acts will contest the composition of the people and who constitutes it. However, while writing their own script, activist citizens still follow an *arche*-script that distinguishes between citizens (insiders), subjects (strangers, outsiders), and abjects (aliens) in the first place. Thus, although their acts draw on incalculability, they still follow a framework that keeps the relations between sameness and difference firmly in place and orients the meaning of politics to boundaries—both in terms of transgression and consolidation.

This suggests that the study of citizenship as an act or practice, rather than as a status, involves a study of what counts as an act as well as who is at liberty to act. It also suggests a study of the relation between boundaries, the meaning of politics, and the question of political community in order to better understand the transformative power of what counts as political action. Within pluralist and agonist political theory, boundaries form the condition of possibility for *politics*. However, politics doubles up with sameness, since politics concerns the boundary set up by a norm, a group, or a constative, that is, anything that marks out a sameness or identity from the different or other. But since the continuance of the possibility of politics must be guaranteed, it follows that political action must result in new boundaries and new communal identities. Boundaries remain as the *arche*-script of political action. And with boundaries, the assumed sameness of the community persists due to it harbouring a common identity or due to it being attuned to a foreigner whose gift consists in its transformative taking. If politics finds its impetus in boundaries, the contestation of boundaries must result in a new closure of identity, community, or political existence for politics to continue to remain possible. Boundaries, it thus seems, appear to form their own paradox, since they on the one hand make politics possible and on the other pose its closure.

THE ANARCHICAL PRINCIPLE OF DEMOCRACY

Central to French philosopher Jacques Rancière's work is a critique of social organisations and orders that present themselves in ways that reduce their corresponding political realities. He refers to images and ideas that present themselves as conclusive whilst neglecting the extensive variations that fall outside their scope of representation as *consensus*. Consensus consists in a certain way of *counting* that manifests a particular image of the community as the "whole of community" and asserts that nothing or no one remains external to the self-characterisation of the community (Rancière 1999). Rancière refers to this as a "distribution of the sensible", that is, a particular "distribution of spaces, times, and forms of activity" that is largely unquestioned and sets up its own frames and possibilities for action, power, measures of inclusion, and ways of life (Rancière 2004, 12).

The taken-for-granted character of the consensus community appears to drain social life of politics. Nevertheless, the consensus community is political to its core, since it "is structurally divided, not between divergent interest groups and opinions, but divided in relation to itself" (Rancière 2010, 188). Assuming a mere part of the community as encompassing the whole, the consensus community erases its intensely political foundations from its formal presentation. It forms a "police" community, that is, a system dedicated to legitimising and maintaining a specific societal distribution of roles and positions within that society (Rancière 1999).

Rancière's identification of the consensus community as our contemporary form for social organisation has effects for the meaning of politics, the political subject, and democracy, which all emerge as highly radical categories that draw on an anarchical core. *The political* consists in the identification of a *wrong* that addresses the gap between how the community represents itself and the unreality of that representation. This is articulated as a *dissensus* or *disagreement* that reveals the contingency of the distribution of the sensible—including its closures and univocal representations of social reality—to itself. Disagreement thereby discloses the political character of the distribution of the sensible and makes it open to dispute (Rancière 1999, 2010).

The political subject is the one articulating or staging such disagreement. This is not done through merely pointing at an injustice that needs to be confronted or an inequality that needs to be levelled out. Instead, dissensus consists in the very contestation of what makes up the community. The emergence of a *sans-part* or "a part of those who have no part" points at the failures of the police logic to represent all of those who fall under its organising power. An assumption of anyone's equality with everyone forms the driving force behind this notion of the political and the political subject.

With a notion of politics as predicated on an assumption of equality as a given, Rancière rejects the idea that equality is something gained through inclusion or that it is a goal governments and societies can strive to reach. Anyone who "assumes inequality and proposes to reduce it, can only succeed in setting up a hierarchy of inequalities, a hierarchy of priorities, a hierarchy of intelligences—and will reproduce inequality ad infinitum" (Rancière 2007, 52). Instead, Rancière forwards the assumption of equality as the starting point of politics. Equality can be staged, no matter the factual inequalities and exclusions at hand. Through claiming themselves as equals, the *sans-part* reveals the community as a contestable order of inclusions and exclusions and introduces a disagreement between the police logic and the egalitarian logic (Rancière 1999). The *sans-part* emerges as a political subject precisely through defying the boundaries of existing categories, through claiming as given the equality that they lack, that is, through acting and speaking *as if* they were the political subjects they are not recognized as. Thus, political subjectification entails not the assertion of an identity. On the contrary, it is a *disidentification*; a denial of identities ascribed by the prevailing order.

Rancière effectively circumvents the relations of inequality inherent in the sovereign decision on inclusion and reverses the image of inclusion as something that happens from inside out. A quantitative form of inclusion seen in practices of recognition serves merely to expand the reach of the common but avoids addressing the particular police order that gave rise to the exclusions in the first place (Bingham and Biesta, 2010, 82–83). Instead, assuming equality as a given makes possible a qualitative, rather than numerical, change of the community through positing a different logic by which the count of the community is carried out.

Democracy draws its very power from this radical understanding of equality. Democracy

> means this astounding principle: those who rule do it on the grounds that there is no reason why some persons should rule over the others, except the fact that there is no reason. This is the anarchical principle of democracy, which is the disjunctive junction of power and demos. The paradox is that this anarchical principle of democracy turns out to be the only ground for the existence of something like a political community and a political power. (Rancière 2009, 277)

In a democracy, there is no ground for why some should be more qualified to rule than others. On the contrary, the central premise of democracy is the contingency of the constellation by which power is distributed, since this allows for the rule of anyone. Thus, democracy is not a certain organisation of polticial life. Democracy *is* politics. Democracy is the realisation of the lacking foundation of power and radical heterogeneity of the *demos*. The democratic principle consists in that those who rule are not qualified to rule by any particular characteristic; they do not rule because they are wiser, more apt, or good or bad persons. On the contrary, the power of the people is the power of those who have no qualification (ibid.). Herein lies the disjunctive junction of the power and the *demos*. On the one hand, there is the power consolidated in the social relations of the police. On the other, there is the power of the *demos*; "the power of whomever" (ibid.). These two powers are disjunctive in that the first accepts a dissymmetry at the ground of the community while the second does not. The junction of the two consists in them both relying on the democratic principle; that is, the insight that "their power rests on its own absence of legitimacy" (ibid. 277–78).

The anarchical principle of democracy attaches a radically transformative power to politics. Markus Gunneflo and Niklas Selberg give an instructive example of the political subjectification of the *sans-part* in their study of the network Papperslösa Stockholm. Through studying undocumented migrant workers' demands to enjoy political inclusion and legal protection through membership of a labour union, they identify an example of Rancière's notion of disagreement. When demanding rights and inclusion at a hearing in the Swedish parliament, the organisation Papperslösa Stockholm, consisting of undocumented migrant workers, challenged "the specific place assigned to undocumented migrants in the … consensus understanding of undocumented migrants" (Gunneflo and Selberg 2010, 13). The undocumented migrants in Papperslösa Stockholm appear as the political subjects articulated by Rancière on especially two points. First, they express their opinions and demands regarding an amendment legislation concerning the regulation of immigration of foreign labour, even though not formally qualified to do so. They act on the presupposition of their equality and challenge the place normally assigned to undocumented migrants. Second, they do so in a forum constituted by a national parliament. Thereby they establish themselves as "a part of those who have no part" and reconfigure a "shared field of experience through the introduction of a wrong" (ibid. 14). This discloses the emancipatory potential in practices that could appear as taking place at the margins of politics; the disagreement expressed through the claim for protection of undocumented migrant workers extends to a disagreement on a system of political communities that enables us to "distinguish between workers and undocumented migrants in the first place" (ibid. 21).

With Rancière, a space opens up in which the *sans-part* emerge as a taking foreigner, who while taking the rights she practically lacks also gives something to the community when exposing it to its own divisions and poses an augmentation of its image. Not only do the undocumented migrants claim a right, they also pose a disagreement concerning the very setup of the community. Indeed, Papperslösa Stockholm enacts another community when they confront the prevailing order with an egalitarian logic that suggests that each and everyone can assume whatever places within that order. Yet, if disagreement should not simply pass ignored, it seems that the giving entailed in disagreement forms a two-way route. The latter begs the question of what is the source of Rancière's notion of equality. Is it found in, for example, nature, history, or reason, and can it be ascertained that everyone has access to this?

It could be claimed that the reason the police can apprehend the claimed equality at all is "because the inegalitarian order itself always presupposes the equality of individuals as speaking beings in its functioning" (Corcoran 2010, 9). If so, then Rancière's assumption of equality as a given appears to draw an image of political subjectivity not very different from the rational subject participating in deliberative discourse posed by Jürgen Habermas. Habermas's deliberating subject is, through a presupposition of equality, taken to be capable of abstracting itself from the particularities of its social position and engaging in rational deliberation. However, in comparison, the difference between Rancière's emphasis on "intelligence" and Habermas's rational deliberation is apparent.

Rancière is careful to articulate the need *to assume* equality *as* a given, instead of saying that equality *is* given. Equality is an assumption, a point of departure, which offers a possibility to redescribe the form and content of the community as well as the meaning and location of political subjectivity. The principle of equality is necessary for politics, but not as a founding ontological principle of politics nor as a goal to which politics strive. Equality is not a rule known in advance (Markell 2003); rather, it is an "empty universalism"

(Žižek 1999, 244) that draws its emancipatory and transformative power from the possibility of communication.

Rancière's work on pedagogy in *The Ignorant Schoolmaster* is instructive. The book addresses how *explication* as a teaching method establishes hierarchical relations between the master and her students, that is, between different intelligences. Through explication, understanding forms a process of stultification in which an active superior knowledge guides a passive inferior one. Intellectual *emancipation*, instead, implies that the master's task is to reveal the student's intelligence and capabilities to the student herself and demand that the student becomes conscious of her own equality with anyone and everyone. If explication relies on inequality, that is, on the assumed inferiority of the minds addressed, the very possibility of communication relies instead on the assumption of the equality of intelligences (Rancière 1991). Through the will to communicate, the "human intelligence employs all its art to making itself understood and to understanding what the neighboring intelligence is signifying" (ibid. 62). Thus, the very possibility of communication rests on the assumption that one intelligence is equal to the other; indeed, "the equality of intelligence is the common bond of humanity" (ibid. 73).

In *Disagreement*, the political implications of the possibility of communication are made more apparent. Whereas the assumption of the equality of intelligences in *The Ignorant Schoolmaster* assures the individual of her capacity to self-empowerment and emancipation, the claim to be part of a communicative regime in *Disagreement* functions as a way to undermine social authority. Two intelligences understand each other because they both take "part in the same community of speaking beings and so [are], in this sense, … equal" (Rancière 1999, 49). This opens the way for two contrary readings of what is at stake in relationships of social inequality, something that Rancière investigates through the meaning of giving and obeying orders.

In the first reading, an order given by a superior would be followed by an inferior as a consequence of their unequal relationship. The superior imposes his language on the inferior; he manifests a particular distribution of the sensible with entrenched internal divisions and presents this as the agreed form for communal coexistence (ibid. 45–49). Possibilities for resistance here rest in formulating a conflicting idiom of power. However, this upholds the divisions and polarisations inherent to the sensible as presented by the superior, although with a potential shift of power relations.

In the second reading, the reason why an order is followed lies in the *possibility* of communication between the superior and the inferior. Despite their social inequalities, they both take part in the same community of speaking beings and are in this sense equal. The inferior rejects the divisions established by the superior's recourse to a specific language. Instead, the inferior insists on their capacity to understand the superior and therefore also on their equality as speaking beings. That social inequalities at all are possible, this reading suggests, is because of the equality of speaking beings (ibid.). This paradoxical nature of communication cuts at the logic of the consensus order and directs our attention to an assumption of equality buried at its heart, which, if brought to light, subverts the distribution of power and ranks of the common. This understanding provides for another kind of resistance, which is "upsetting, in the proper sense of the term" (ibid. 49). When stating, "We are right to argue for our rights and to posit the existence of a common world of argument" (ibid. 53), the ones enacting equality *set up* another distribution of political and legal boundaries. They challenge the police by presenting a recalculation of whom to count as included in the common. The police forms "an order of the visible and the sayable that sees that a particular activity is visible and another is not, that this speech is understood as discourse and another as noise" (ibid.). The dissensus posed by the undocumented migrant workers of Papperslösa Stockholm consists in the claim to produce *voice* rather than *noise*, that is, of expressing *logos* instead of *phone*.

This distinction between *phone* and *logos* is the basis of politics: "Politics exists because the logos is never simply speech, because it is always indissolubly the *account* that is made of this speech: the account by which a sonorous emission is understood as speech, capable of enunciating what is just, whereas some other emission is merely perceived as a noise signalling pleasure or pain, consent or revolt" (ibid. 22–23). Hence, some speech is appreciated as voice; some heard as mere noise. If perceived as noise, or *phone*, the action of Papperslösa Stockholm will go unnoticed; the police has failed to hear how they articulate themselves as *a part* of those who have no part. When heard as voice, Papperslösa Stockholm makes an account of themselves as speaking within the realm of *logos*, or, more accurately put, they make an account of *logos* as inclusive of their speech. This means not that the realm of *logos* is merely expanded to encompass also the claims of the demonstrating undocumented migrant workers. Such a numerical inclusion would not form a dissensus. Instead, dissensus signifies the positing of a disagreement concerning what account of *logos* constitutes the *logos* of the community (ibid. 37–40). Two logics of accounting for the *logos* of the community are set against each other, but the fact that the police logic hears, or is forced to hear, the claim of Papperslösa Stockholm as an articulation of voice, rather than mere noise, means that the police logic is already disqualified as the all-encompassing count of the community. The politics emanating from the assumption of everyone as equal speaking beings hence serves to transform the community in total and pose a community different from the former consensus.

Here one sees how the resemblance between Habermas and Rancière begins to fade. More than Habermas's orientation of politics towards consensus, their difference lies in the relation between reason and equality. While not duly recognising this, May points out how Habermas's account of reason is grounded in people regarding each other as equally intelligent and with an equal part in articulating the best possible norms of the polity (May 2010, 14–15). To May, this parallels how Rancière finds that "reason begins when discourses organized with the goal of being right cease, begins where equality is recognized: ... an equality in act, verified, at each step by those marchers who, in their constant attention to themselves and in their endless revolving around the truth, find the right sentences to make themselves understood by others" (Rancière, quoted in ibid. 14). May takes this quote from *The Ignorant Schoolmaster*, and according to the realm of intellectual emancipation, his conclusion is correct; *reason* begins when equality is assumed. That anything can be communicated is based on the assumption that there is an equality of intelligences. Hence, the very possibility for a community of speaking beings rests on the assumption of an equality of intelligences.[1]

A realm of *logos* would thus emanate from the assumed equality of intelligences wherefore it would be correct to say that reason begins with equality. In the question of *political emancipation*, the situation is, however, the reverse. The assumption of equality gains force *at the end* of a given realm of *logos*. For Habermas, communicative reason would guarantee against claims of inequality, or, put differently, reason begins where equality ends. For Rancière, the opposite is the case; equality begins where a given realm of *logos* ends. There, equality assumes its upsetting force and constitutes the guarantee against a totalising account of *logos*. When it concerns politics, equality becomes crucial first at the end of a particular community of speech. Despite being predicated on the same feature—an assumption that corresponds little to the factual situation—equality functions differently in intellectual and political emancipation respectively. For intellectual emancipation, equality is a point of departure that opens up a field of possibilities. For political emancipation, equality finds its gravity at the limits of a given sensible. Thus, contrary to intellectual emancipation, the assumption of equality necessary for politics requires a *boundary* in order for emancipation to be possible.

To Rancière, the source of equality therefore differs depending on the realm. For intellectual emancipation, equality draws its transformative strength from the *possibility of communication* between intelligences and the assumption of intellectual equality this possibility rests on. Once the equality of intelligences is realised, reason knows no boundaries. Or, just as equality is synonymous with intelligence, reason is synonymous with will (Rancière 1991, 71, 73).

Political emancipation also draws its power from the communicative regime, but political speech, contrary to intellectual emancipation, is dependent on being heard by the police. Lacking pre-defined content, equality is empty in itself and can be appropriated by *any* intelligence and work as a vehicle for action. It forms an act of subjectification that undoes the supposedly natural order of the sensible. Still, in the case of politics, equality finds its source in a shared *logos* only as long as such sharing is both articulated and recognised. This, however, means that assumption of equality put to work in political emancipation requires exactly what the equality of intellectual emancipation undoes—boundaries.

What emerges from Rancière's anarchical principle of democracy so far is an understanding of politics as a radically transformative undertaking that finds its impetus in the assumption of equality as a given. Boundaries animate politics, not simply in terms of pushing them but, in revealing their contingency and altering, turning the question of political community into disagreement on what can be grasped as expressions of *logos*. Yet, democracy, Rancière says, consists in the disjunctive junction of the police logic and the egalitarian logic of the *demos*. But how is this disjunctive junction possible and what is it premised on?

A COMMUNITY OF EQUALS?

Balibar emphasises how the political claim to the rights one is denied "describes a process which started with resistance and ends in the actual exercise of a 'constituent power'" (Balibar 2002, 167). Following Marx, the notion of participation encompasses the notion of the community, as the content of political rights is participation, and participation can only be exercised in the community (Marx 1978). But when the boundaries of the community challenged by disagreement are redrawn, a new consensus is established, with new ways of counting on the common and with new internal divisions and relationships of inclusions and exclusions. In this, the newly realised substitutability and contingency of the community is concealed in the new consensus, while the political action of the *sans-part* is consumed into the new police.

Rancière's understanding of equality suggests the "substitutability", or the equivalence, of the same and the other and a notion of sameness that disregards any relations of otherness, except for the one between the *part* and the *sans-part* (Rancière 2009, 278). Is this not a sameness that avoids presupposing difference? Rancière says almost as much when he articulates sameness as *indifference to difference*. While paramount to Rancière's idea of politics, this understanding of sameness also effectively prevents any sustained community of equals.

In the disjunctive junction between the police logic and the egalitarian logic that underlies the anarchical principle of democracy, the police logic consists in the lacking ground of the hierarchies and power relations of the police in a democratic regime. The egalitarian logic consists in incalculability of the *demos* and the power of the *demos* as the power of whomever. The lacking legitimacy of democratic power is crucial because this vouches for the substitutability of power. At the moment when these two logics meet, the indifference to difference inherent in the egalitarian logic can be realised. At that moment, the *demos* takes stage, by challenging "parts, places, and competences by linking a peculiar wrong done to a peculiar group with the wrong done to anyone by the police distribution" (ibid.). But that moment is passing; the people only act *as if* incalculable, suggesting that at some moment, the people must cease to be and the community must resort to counting.

The reason for this becomes apparent in Rancière's discussion of Derrida's notion of "democracy to come". This emphasises the impossibility of identifying democracy with a specific form of government and is in that sense similar to Rancière's insistence on each police order as a miscount. However, whereas Rancière's anarchical principle of democracy underscores democracy as the supplement that makes politics possible, Derrida, on Rancière's reading, understands "democracy-to-come" as a "supplement *to* politics" (ibid. 278). Democracy-to-come should not be understood as something taking place in the future, neither is the supplement implied by the "to-come" to be added to democracy from the outside. Instead, Derrida's notion signifies "democracy *as* democracy-to-come" (ibid. 275), meaning that democracy implies an absolute otherness, both asymmetric and incalculable to the self. Thus, Derrida completely dismisses the possibility of *any* form of brotherhood or similarity.

For Rancière, this means also the dismissal of any form of substitutability. Refraining from the idea of substitutability, of indifference to difference, or of the equivalence of same and other, Derrida also refrains from attempting to make sense of the *demos* of democracy, according to Rancière (ibid. 276–78). He sees Derrida replacing the *demos* with *hospes* as the subject of his "democracy-to-come" when organising the relation to the other through the notion of unconditional hospitality. The notion of *hospes* is also a mark of distance. Surely, it characterises an affinity that reaches across communal boundaries, but it leaves no room for reciprocity. *Hospes* as the subject of democracy-to-come posits the other as an absolute other, an other who can never be gained complete access to and who hence cannot be substituted or the subject of substitution. To Rancière, the absolute otherness of Derrida's *hospes* effectively rules out any possibilities of politics. A duty of unconditional hospitality to any newcomer inscribes *hospes* as an *ethical* subject, leaving democracy-to-come void of politics. The absolute otherness implied by Derrida is the reason why Rancière characterises democracy-to-come as a supplement *to* politics, rather than something that falls within its rationality. Hence, when Derrida conjures democracy as something "to come", which is the same as saying that it is "not being here", Rancière can claim Derrida to rule out the "aesthetical dimension" of politics, namely the staging of a dissensus (ibid. 282–83).

To Rancière, instead, the *demos* must be the locus of politics and constitutes both the object and subject of political contestation. The political subject emerges in the process of articulating and staging the *demos*, a process that depends exactly on the substitutability of what the police order conceives of as the people. This process puts "two worlds, two heterogeneous logics on the same stage" (ibid. 276). In this, "brotherhood means a certain equivalence, a certain substitutability" (ibid. 279–80), Rancière suggests, and he poses a "brotherhood", or sameness, that cuts through any identity claims or subject positions and is therefore indifferent to difference. The question of political community, then, consists not simply in opposing one count to another, whilst claiming one as correct and just and the other as incorrect and unjust. It is not even a question of opposing two different logics for calculation. Instead, it means to enact the incalculability of equality in the face of the count of the consensus community.

This turns the *demos* into an event, augmenting, contesting, but also transitory. The only political community of equals that can be is momentary. The power and difference of the *demos* is "vanishing" and "tends to be unrelentingly annulled by the oligarchic functioning of the institutions" (ibid. 278). Alain Badiou addresses how the community of equals is excluded from Rancière's theory, since this would demand a social bond without a master. The community of equals would close the interval between *logos* and *phone*, which is precisely the interval inhabited by Rancière's political subjects. Hence, the egalitarian moment is always a moment "declared and delineated" in which "equality is conceived in its intrinsic bond with inequality" (Badiou 2005, 110–11, at 111). Equality is intrinsically bound to inequality because politics requires that equality is enacted in a situation of inequality, within a wrong (ibid. 115). To Badiou, Rancière's problem concerning the political community of equals is, perhaps, that he remains a democrat (ibid. 120). Despite his critique of police mastery as an imposture, he does not do away with mastery *per se*. Instead he maintains "a pure empty mark of mastery, whose vertical absence provides the foundation for the horizontal bond" (ibid. 109–11, at 110). Herein lies the core of the problem which is also the core of the political—that democracy as generally understood requires a count. Contrary to the rejection of an intellectual master, the surges of political coexistence of egalitarianism feed off mastery. To persist, each transgression of the consensus count needs to pose a new count. Although with possible newly won insights of its contingency and contestability, the new count is still bound to constitute a miscount of its own. Thus, to Badiou's conclusion that equality doubles up with inequality, I would like to add that *sameness is intrinsically bound up with difference* for Rancière. Although the orientation to difference of the egalitarian logic is indifference, there is nothing indifferent to the bond itself. On the contrary, the intrinsic bond of sameness in the face of difference is what makes politics possible. For the egalitarian logic to make sense, and for it to hold its upsetting political capacity, it cannot posit *all* as same in enduring terms. The equality of anyone with everyone is always enacted in the face of inequality, while the indifference to difference is political only when asserted in disagreement with orders established around the same-different nexus. Hence, precisely as an absolute otherness would be ethical for Rancière, absolute equality would also be so. A community of equals would mean the end of politics.

The anarchical principle of democracy posits the question of political community so that both are required—the police logic and the egalitarian logic at once. Democracy's disjunctive junction between a police logic and an egalitarian logic is premised on the assumption of a notion of sameness that presupposes difference *and* the radical similarity expressed in the indifference to difference. The incalculability of the egalitarian logic can meet the police count first when debunking the claimed legitimacy of the latter and exposing this to radical equivalence.[2] Here, Rancière withdraws from the assumption of equality as the *sole* principle of politics. Instead, to sustain his meaning of politics, the disagreement between the counting of some as equal and the incalculability of the equality of anyone with everyone is conditioned *both* on the assumption of equality as a given *and* a notion of sameness that presupposes difference.

RANCIÈRE'S REMAINDERS

Thus, agonist political theory locates the meaning of the political in the transgression of the interval between different heterogeneous logics of the community. Hence, their notion of the political community can be inhabited of, it seems, potentially anyone. At the same time, the community itself is transitory and eventful to its character, tied to the enactments of its subjects. It is also tied to the transformative enactment of the equality of *logos*, which alters the understanding of what can be understood as voice. Would, then, anything pass as *logos*, or is there something that cannot escape being characterised as *phone*?

In an enthralling reading of Sophocles' *Antigone*, Honig shows that the distinction between *logos* and *phone* is not clear-cut. Instead, the lack of purity in language displaces language in the interval between *logos* and *phone*.

Parody, mimicry, and citation are not just a matter of words but involve also intonation and inflection. Thus, they are instances of "*phone*'s eruptions into *logos*" (Honig 2013, 142). Honig follows Rancière's call to work the intervals between two identities. Thus, she sees the mourning Antigone as engaged in the "interstitial rhetorical devices … [that] allow democratic actors to reconfigure distributions of public and private, universal and particular, mortal and citizen. Such reconfiguration is always needed because even the universal, Rancière says, 'is incessantly privatized by police logic'" (ibid. 145–46). Honig finds room for Antigone's dirge in the interval between *phone* and *logos*. Inhabiting this interval, Antigone's lament upsets the oppositional logic between identities: "Her dirge neither conforms to the expected form nor simply violates it. Instead, she parodies, mimics, lampoons, and cites the stories, figures, and speech of the powerful, insinuating her view into their discourse, not absolutizing them, … nor losing hold of her capacity to make sense" (ibid. 146). Hence, her relation to language is one of conspiracy by which she manages to make a new kind of sense (ibid.).

With Honig's Antigone in mind, the question seems not to be if there is any kind of *phone* that is prevented from entering into *logos*, but rather if there is anything that *risks being omitted from them both*. For the displaced, it appears that such a risk cannot be precluded. For the potentially hidden, for the possibly clandestine, what are the options to stage equality? What if there is a part of those who have no part that indeed relies on their own silence? What if keeping quiet regarding their location, their affiliations, and their undertakings is essential for the relative safety of themselves and their loved ones? It seems as if the potential silence of the clandestine would fall outside of both *logos* and *phone*. What, then, does the problem of the silence of the displaced reveal about Rancière's notion of politics?

In order to avoid misconceptions, it should be underlined that to say that the displaced might remain silent certainly does not imply her *inability* to articulate herself as an equal speaking being. Rather, it concerns whether or not she is *at liberty* to speak. As Quentin Skinner emphasises in his reading of Isaiah Berlin, the "lack of freedom must be categorically distinguished from the lack of ability" (Skinner 2002, 245). Berlin insists on the absence of interference as the mark of liberty. Being unfree means that one is prevented by others—intentionally or not—"from doing what [one] could otherwise do" (Berlin quoted in Skinner 2002, 244). The "deliberate interference of other human beings in the area [in] which [one] could otherwise act" would thus be the cause from which loss of freedom stems according to Berlin (Skinner 2002, 244). To this, Skinner adds how also non-human agencies, as for example the forces of nature, might be the cause of disempowerment and rejects consequently the requirement of *deliberate* interference. Skinner goes further and highlights a notion of liberty not engaged by Berlin's largely Hobbesian account of negative liberty. Relatively forgotten, Skinner contends, is how Hobbes's contemporary insisted on a notion of freedom that evaded the dependence on any royal power. On their account, freedom is not only limited "by actual interference or the threat of it, but also by the mere knowledge that we are living in dependence on the goodwill of others" (ibid. 247). Thereby, dependence is seen to undermine liberty. Absence of interference is hence not sufficient for conceptualising negative liberty; instead absence of dependence needs to be taken into account.

That the displaced in this sense, as in others, is unfree is rather obvious. She is most certainly living in dependence of the goodwill of others. She is also dependent on a certain degree of security in order for her to articulate her political existence. But it seems that also the *sans-part* is unfree in the sense articulated by Skinner. Not only because of the lack of security that many of the most precarious individuals can be suspected to experience. In addition, even when they have the possibility to enact themselves as equal, the *sans-part* is dependent on the responsiveness of the consensus community. Only when *apprehended* as articulating an account of *logos* will the interrupting subjects pose the substitutability of the *demos*. When the community fails to show such responsiveness, the interruption will merely be perceived as background noise. A great reward with Rancière's notion of equality is how its enactment is saturated also with liberation. However, this is precluded from his account of political subjectification when taking the republican view of freedom as described by Skinner into account. When Rancière inscribes the staging of equality in the domain of *logos* and language, he constrains that staging through the need for the *sans-part* to be perceived as articulating voice rather than noise. Rather than a staging of equality that is also a staging of liberty that disregards relations of domination and dependence, Rancière's equality emerges as trapped in a relation where the lack of freedom becomes a central predicament.

For the displaced, her lack of freedom prevents her from enacting equality altogether, at least not without risking a price perhaps too high to pay. In the case of the *sans-part*, say, in the form of the migrant workers of Papperslösa Stockholm, they are dependent on the goodwill of those allowing them to be heard. Further, their enactment is dependent on the benevolence of those already included. In order to quickly absent themselves from the open they momentarily step out into to stage their equality, the paperless workers depend on people with resident permits who can help them with the logistical matters necessary for a staging to be made (Sigvardsdotter 2012, 122–23).[3]

At first glance, it seems like Honig opens for the possibility to escape a strict dependence on *logos*, as well as on responsiveness; the "shout-it-from-the-rooftops-character" of Antigone seems not to wait for people to listen. Rather, she appears to impose herself on language, acting in conspiracy with language as language is in conspiracy with her. However, the problem of dependence becomes apparent when attention is directed to her sister, Ismene. In a remarkable way, Honig invites the often disregarded Ismene to take centre stage in her reading of the play. Honig highlights how, before Antigone's burial and lamentation of her brother, another burial takes place. This is carried out in the dark of the night, without revealing the one responsible for the act. Honig attributes this first burial of Polynieces to Ismene. Through the sisters' sororal conspiracy, Honig sees the contours of the actions of Ismene take form (see chapters 5 and 6 in Honig 2013). But it is Antigone who brings them forth in a *sotto voce* conspiracy with her sister. Alone, even Ismene's *explicit* articulation of her part in the burial of her brother goes unnoticed, by Creon as well as by centuries of readers. Even in Honig's reading, the quiet Ismene seems quite dependent on her outspoken sister. Had it not been for the latter, who does not shy away from burying her brother in broad daylight while loudly lamenting her loss, the covert action of Ismene's nightly burial would perhaps have remained in the dark.

While Honig writes of how Antigone's two laws of equality and singularity do not give priority to either *logos* or *phone*, she also ascribes a certain awareness to Antigone. Tracing the references of her claims and of her mourning in other plays of the same era, Honig draws an Antigone who appears highly capable of surveying the rules from which her plotting and conspiracy take off.[4] Indeed, she might inaugurate new publics; Antigone's laws are the "twinned, agonistic commitments of democracy" (ibid. 123). Yet, one wonders, are these not the laws of Ismene too? And perhaps even more so, considering how the equality she stealthily enacts does not conspire with language or the knowledge of the Athenian audience, though from the act remaining in the dark, no new publics seem to unfold.

Rancière's concept of equality reintroduces thereby—paradoxically—both the requirements of liberty and of brotherhood, or sorority on Honig's note, through its reliance on language. The possibility of staging equality is always already circumscribed by a lack of liberty due to the dependence on the responsiveness of the already included. Equally, Rancière's notion of equality must also assert a delineated sameness alongside the infinity of the claim to brotherhood or similarity, in order to not preclude future politics. The insistence on the equality of speaking beings that give an account of *logos* creates its own remainders. For, alongside his community of speaking beings, a part of those who have no part yet again appears to take form, or at least a part of those that, in order to safeguard the lives of themselves and their families, need to refrain from any explicit articulation of their equality and of themselves as an already existing part of the common. Thereby, despite eschewing inclusion as a "mediation of inequalities", Rancière inscribes political subjectification as a mediation of unfreedom through his emphasis on political voice.

Hence, Rancière's disagreement draws on its own police logic through the demand on the faculty of language—no matter what is understood as such—for staging a wrong or articulating disagreement. The mediation of unfreedom at work in the necessary responsiveness of the consensus community discharges political subjectification of its egalitarian presumptions. Furthermore, when the situation of the non-included demands her silence—or a stealthy act in the dark—there appears to be no entry point at all for staging an equality by which anyone and everyone are assumed as equal. Must freedom then not also be assumed, one wonders, in order for Rancière's notion of politics to be put to work?

CONCLUSION

Agonist political theory redirects the energies of the question of political community from concerns of who and what should be included, to revolving around the political acts through which the community is staged. However, agonist politics is animated by and premised on boundaries. These boundaries tie the reach of the political community to the limits of how our world is rationally ordered, meaning that agonist politics seek to expose what previously has been heard as *phone*, or mere noise, as *logos*, or political voice. Now, if agonist politics is premised on boundaries, and such boundaries ultimately are found in the *logos* of the community, it follows that agonistic politics seeking to push, transgress, and transform boundaries are premised on a certain visibility. If a political act should inaugurate new publics, it is suggested, that act needs to be *public*. For agonist political theorists, then, the question of political community as a question of how politics, equality, and belonging can be enacted must also be accompanied by the question of *by what capacity* political acts can be staged. For one could presume that those who have the most to win from claiming their rights to have rights are also the ones with the most to lose from the public exposure of political enactment.

Thus, while a notion of politics as consisting in a re-enactment of the *logos* by which identities, categories, places, and positions are distributed in society allows for formally not-included to enter as the protagonists of the political stage, it also closes that same stage to those in the most vulnerable positions in said society. Redrawing the limits of community to include what lies beyond a given realm of *logos* means to still rely on the faculty of speech and the assumption that one is at liberty to speak. Stateless groups, now in clandestine existence, again emerge as a symptom of the obsession with visibility, voice, and public appearance of our time.

The agonist political subject has certainly altered the meaning of the political community. She does not wait for a right to be handed her. She acts, free from interference, and alters the rationale of the community accordingly. Yet, with a politics structured around boundaries and with its source in the equality of *logos* or political speech, she acts in dependence, meaning that playing a part in the political community is first of all dependent on her being at liberty to do so.

NOTES

1. In *The Ignorant Schoolmaster*, Rancière in fact articulates equality to be synonymous with intelligence (Rancière 1991, 73).
2. Further, the anarchical principle of democracy consists in the rule by anyone for no particular reason. Thus, the formalised substitutability inherent to democracy gains precedence over the non-formalised incalculability of the *demos*. One could perhaps see Rancière's position as a negative formalism, making his theory and Habermas's formalism closer than either of them would like to admit.
3. Sigvardsdotter followed the network Papperslösa Stockholm during several years of demonstrations outside the building of the Swedish parliament before they were admitted entry to speak to the parliamentarians. She describes how her hopes to speak to some of the paperless after the demonstrations were futile, as they quickly absented themselves from the scene, leaving people with resident permits to pack up the equipment. While the latter helped provide the scene, the paperless again put on the "invisibility cloaks of borrowed semblance" (Sigvardsdotter 2012, 122–23).
4. I owe this point to Ted Svensson.

Chapter 3

Space

The previous two chapters have shown how statelessness as a symptom of modern politics is symptomatic of how boundaries animate the meaning of the political and of the political community. Boundaries were seen to mark out membership, determined on the basis of a particular identity that made the political community conditioned on a notion of sameness that presupposes difference. Boundaries emerged as the boundaries of a particular regime of *logos*, marking out a distribution of possible positions beings can inhabit in relation to the community. Boundaries also came forth as the primary concern of politics; they formed a ground for understanding politics as contestation, inauguration, and reidentification. But, by forming a premise for politics, boundaries were also seen to set certain bounds on what it means to act politically, through the expectation of political action to be visible, publicly enacted, and articulated. By so doing, the boundaries that would spark political enactment also installed a strict separation—between those at liberty to speak and those who are not.

Alongside statelessness as symptomatic of how boundaries mark out a collective identity, or determine who belongs to a particular rationale, statelessness is also—and most obviously—a symptom of boundaries in space. Statelessness as a symptom is symptomatic of a world compartmentalised into territories and the concomitant understanding that each population has a territory of its own. But the spatialities that statelessness is symptomatic of reach further and describe spaces in which some can appear and move freely while others cannot. Spaces of fundamental inequality emerge anyplace within a given territory and put to work boundaries productive of externalities internal to the state.

Political theory has increasingly sought to engage the territorial state and its exclusions through addressing the issues of space and the spatiality of identity, community, and belonging. Thereby, the question of political community is oriented towards the issue of *where* political community can be seen to emerge or take place. This chapter will explore some of these attempts, while also pointing at how the spatialities they describe themselves remain within a realm of representational thinking and are thereby premised on a notion of sameness that presupposes difference. Thus, it will be seen how the exclusive spatiality expressed in territory is replaced by a space of sameness that grounds power and control in contemporary politics.

SPACE AND THE STATE

With his notion of the "territorial trap", John Agnew gave a seminal critique of how state sovereignty is supposedly found and contained within territorial boundaries. Agnew questioned the assumptions that space is primarily territorial, that is, that space can be described according to a geographical scale conditioned on state boundaries. It becomes particularly problematic to emphasise the territorial sovereign state in international relations theory, Agnew argues, since the latter posits a dehistoricised and decontextualized ideal-type of the state as its central object of inquiry. Further, the assumption in international relations theory that the territorial state can be understood through distinctions between the national and the international, or the domestic and the foreign, obscures how relations and processes of political and economic life develop and operate on different scales in society. Lastly, to sustain the territorial state as the centre of analysis risks seeing territory as an already present 'container' in which societal relations can develop. The latter suggests that territory is a prepolitical entity that exists prior to the state and would rather be circumscribed by the reach of the nation (Agnew 1994).

Returning to the earlier discussion on the legitimacy of the people, one can here see how territory steps in to ground the boundaries of the *demos*. As argued by Ochoa Espejo, this goes two ways. While a number of democratic theorists picture an, in principle, unbounded *demos*, this will in practice be delimited according to prepolitically established territorial boundaries, meaning that the boundaries of the *demos* become a function of territory. On the other side, she sees theorists arguing for self-determination as a right exercised over a given territory. This, however, means that a well-defined people is taken to justify exclusionary claims to a territory, turning territory into a function of peoplehood (Espejo 2014, 5–7). In both cases, however, territory emerge as the central spatiality of the political community.

Drawing on Henri Lefebvre, Agnew highlights how the above merging of the territorial state with the community must not necessarily be understood as an "intellectual illusion". What is illusory is instead the self-proclaimed unity and homogeneity of the territorial state. In order to uphold the state, spatial unity and internal homogeneity are "*taken for granted* as a 'reality' of social life in general" (Agnew 1994, 71). In this, the territorial state becomes the means of managing difference, by providing itself as the vessel that harbours a particular identity and separates this from other societal identities—from other delineated communities in the geographical grid of bounded states. Agnew prompts his readers to escape this trap, which he holds as circular and cumulative, and responses to his call can be found in a variety of disciplines, not least in political geography. There, attempts to engage with the boundaries of the territory and their historical and sociological formation, as well as with the problem of a territorial unity that seems to function as an existential completion of the community, have posed serious challenges to a notion of space as simply a container of the unfolding of history (Gregory 2004; Jones 2009).

Territory connotes but one account of space and practices related to it. Space in itself is vast, both conceptually and materially speaking. It has been explored as central for the constitution of society (Lefebvre 1991; Soja 1989), as embodied enactment in contrast to the textual representations and significations at work in discursive identity formation (Ingold 1993, 1995), as inherently relational (Massey 2005), and as an intimate realm of lived experience in the form of place (Casey 1993, 1997). Indeed, place is not really *an aspect* of space according to Casey's thorough study. Rather, place and the particular cosmology corresponding to a finite place hold a central position in early Western thought and were only later subordinated to the infinity and universality of absolute space, which, in turn, has been subordinated to time. Casey attributes this partly to the infinite divinity of Christian theology and its close relation to the equally universal Newtonian physics, which precludes a finite and experienced notion of place (Casey 1997).

It is precisely because territory is but one way to understand spatial ordering that it needs to be scrutinised closely, Stuart Elden maintains when trying to answer Agnew's call for a less geographical and more historical study of territory. Since state space has historically been inscribed as a delineated territory, Elden stresses the importance of avoiding taking the meaning of territory as self-evident. Too often is territory taken as an unproblematic straightforward concept, and the territorial sovereign state is reified as the necessary model for political organisation. Failing to give an historical account of territory runs the risk of forwarding an anachronistic portrayal of the concept, Elden maintains (Elden 2010a, 2010b, 2013a).

Both technical and legal relations are at work in the production of territory as the central locus for political power. An analysis of territory calls thereby both for a legal perspective, which is crucial to understanding the relations between territory and sovereignty, and a focus on technical issues that is central to understanding what is involved in the definition of territory as bounded space. While boundaries themselves are not a modern invention, the political need to bind land has rendered boundaries a particularly modern meaning that is dependent on a particular notion of space. Therefore, a critical theory of territory must "investigate the quantification of space and the role of calculative mechanisms in the commanding of territory, and the establishment of borders" (Elden 2009, xxvi–xxvii, at xxvii).

Studying the emergence of the territorial state as a spatial unity, Jordan Branch identifies how the practice of cartography has played an essential role in territorialising political authority. Branch underlines how the heterogeneous character of medieval authority was organised around centres of strong authority with unclear boundaries and could also be non-territorial in character, rather than being confined within geometrically described spaces. A delineated territory was allowed to be established as the "structuring concept for political organization and interaction" first with the practice of mapping, that is, of drawing linear boundaries around a continuous and measurable space, Branch argues (Branch 2011, 18). Through appropriations of Ptolemaic mapping, the understanding of the homogeneous territory as "a geometrically measured space, the purely linear character of political boundaries, and the elimination of nonterritorial authorities" thereby served to establish modern political authority and modern statehood in exclusively territorial terms (ibid. 18–19, 27, at 7).

Also Bartelson emphasises cartographical and geographical practices of the late sixteenth and early seventeenth centuries and their centrality for the fusion of authority and territory. With this, Bartelson also draws the bounding of space as a modern phenomenon that was not prescribed by early modern theorists of sovereignty. In contrast, the latter were influenced by pre-modern geometry, in which especially the notion of indivisibility in Proclus's geometry appears as a source for engaging the problem of sovereignty. To Proclus, indivisibility is found in the geometrical image of the point. As to Euclid, the point is what has no parts, but thereby it is also sovereign to things divisible according to Proclus. Hence, "whereas early-modern theorists of the state defined sovereignty in terms of indivisibility, Proclus defined indivisibility in terms of *its* sovereignty over all things divisible" (Bartelson 2014, 18–28, at 23). In both cases, indivisibility suggests neither the unity of what is subordinated to it, nor is it aligned with a homogeneous and delineated territory. Instead, Bartelson maintains how "the quintessentially modern fusion of authority and territory presupposes the possibility of bounding space" (ibid. 28). Further, "through such bounding, space is turned into a *precondition* as well as a *limit* of sovereign authority, thereby being constituted into a *territory* in a recognizably modern sense of this term" (ibid.).

Thus, territory and its emergence pose a "question of the practices that relates power and politics to place" (Elden 2013a, 2013b, 15). Elden emphasises how Foucault foremost understands the strategies of territory to be calculative. This means that the rationality of territory draws on mapping, measuring, and ordering, while it also normalises these strategies and ties them to governance. The delineation and establishment of territory thus align with the same mechanisms that are at work in the production and control of the population; "just as the people become understood as both discrete individuals and their aggregated whole, the land they inhabit is also something that is understood in terms of geometric, rational properties, or 'qualities'" (Elden 2013b, 14). Hence, territory is a "political technology"—a technique by which borders are established and populations formed. And, it can be added, territory and population mirror each other in rendering the people and the space they inhabit calculable.

Together with Neil Brenner, Elden discusses the relation between a bounded space and a bounded people. Drawing on Lefebvre, they underscore how the modern state is productive of the national territory as its physical space (Brenner and Elden 2009, 361–65). Although not providing any exhaustive discussion of the *demos* inhabiting a particular state space, Lefebvre's emphasis on the legitimising grounds for the delineation of the nation as well as its space as something *assumed* hints at an image of the *demos* as a nation. This fits neatly into the fiction of the homogeneity and continuity that describes state space (Lefebvre 1991, 111–12). Brenner and Elden especially attend to how, in Lefebvre, the "concepts of state, space and territory are ineluctably intertwined" so that "each term reciprocally implies the others" (Brenner and Elden 2009, 364). By highlighting this, they want to articulate that any theory of the state must also engage a theory of territory, and that a theory of territory necessitates also an account of the state as well of how territory and the state are productive of space (ibid.).

By their analysis, Brenner and Elden provide a powerful critique of how "the body of the nation [is] translated into political practice" through representations of territory and of their relations to capitalist development (ibid. 365). They discern how territory becomes a means to create an image of the space of the political community as autonomous from the social and political relations productive of it. Further, they identify how territory provides a physical basis upon which the modern state and concomitantly the modern interstate system can be sustained, and how this, as well as its domestic counterpart, is subject to transformation due to continual struggle (ibid. 364). They highlight how Lefebvre's "approach fruitfully suggests that the state's territorial form—state space *as* territory—is in fact quite crucial to the political illusion of transparency" (ibid. 373). Hence, "a genuine escape from the territorial trap hinges upon a reflexive theorization and investigation of territory as precondition, medium, mechanism, and outcome of modern state power" (ibid.). This is certainly an important part in investigating state power since, as noted by Gregory and Branch amongst others, the continuity and unity of a bordered territory has served as a backdrop for the delineation of the people or the nation.

Thus, the bounded territory provides for another stand-in for the lacking legitimacy of the boundaries of the *demos*. With territory, the elusive unity of the community can be reflected in the illusive transparency of space. Both the cosmopolitan notion of world citizenship and Kymlicka's multicultural citizenship can be seen to project the smooth and continuous space of the nation as the *par excellence* space for the political community. This adds another dimension to Kymlicka's insistence on the voluntariness of migration. Having left not only her culture but also her land behind, there simply is *no space* for the migrant to claim her minority rights that would not create holes and rift in the smooth spatial fabric of the community. National minorities, with their historical ties to their land of residence, would on the contrary confirm the land inhabited by the majority culture as invested with a supposedly continuous historical trajectory. Similarly, rather than seeing refugees, stateless, and nomads as the protagonists of cosmopolitan citizenship, Heater emphasises transnational networks, as well as economic and professional relationships, as the driving forces of cosmopolitanism (Heater 2002). Such disregard of the most exposed groups under global politics reflects a worry of disrupting the spatial smoothness of the cosmopolitan ideal. Envisioning a transparent world in which there is place for all cosmopolitan subjects, Heater appears to efface what contests and complicates this vision. Instead, a homogeneous space invested with a presumed universal morality becomes a stand-in for the equally uniform national space. Thus, the continuity of territory serves as a blueprint for the spatiality of the community, whether pluralist or global.

A calculative homogenisation of the people is mirrored in the cartographic calculability of state territory. Of course, state space is not as smooth as suggested by the prevailing order. On the contrary, calculable space has not only homogenising effects. When put to the test of those not upholding formal membership in the community, like undocumented migrants, stateless persons, marginalised groups, or the differentiations within these groups, it becomes obvious how state space is also internally differentiated. These various parts that have a lesser or no part in the community underscore how state space constitutes a differentiated power regime, leaving territory to appear as a transparent organisation of power only vis-à-vis the nation. Further, although homogenisation is upheld by the state system, it unfolds on a larger scale than the individual state.

Lefebvre sees this process taking place on a world scale, on which space is organised as unifying world space with strong centres surrounded by weaker and dominated peripheries, so that "space has much stronger connections with the State than territory once had with the nation" (Lefebvre 2009, 213–14). If Abizadeh's articulation of the democratic boundary problem is read in relation to state space, it becomes apparent how state space transcends the physical basis it might find in the territory. When extending its power beyond its factual borders, and when exercising rule over individuals that are not recognised as democratic subjects, the state produces spaces that organise social relations in particular ways also far beyond the reach of its territory and the *demos* assumed to inhabit this. These spaces can be spaces of control, of domination, of discrimination, and of entrenched inequity. They are all spaces of violence. Maybe they can be seen as boundary spaces—or as border zones in the vocabulary of Balibar—that is, as spaces where the work of state space escalates in its most extreme form. But this would yet again invite the image of state territory as cohesive and transparent and, of most importance, as something relatively *separate* from the spaces of its boundaries.

Lefebvre stresses how space needs to be understood as *produced*, meaning that a study of space must analyse how different forms and possibilities for social and political life relate to particular modes of space production. Neither dead, static, or *a priori* to social relations, space is indistinguishable from knowledge, power, politics, and economy. Space is at the same time presupposed, conceived, and lived. To Lefebvre, the spatiality particular to the modern capitalist state is *abstract space*, with its own particular way of life, understanding, and production. Most crucially, abstract space *appears* homogeneous. Through such perceived homogeneity, differences threatening the institutions of abstract space can be ordered under its rationale. That is, while productive of differences, abstract space orders such differences so that they become "internally acceptable" to the system in which they appear (Lefebvre 1991, 396). Thus, difference is organised according to the same structure to that from which it is different. Or, put differently, both sameness and difference are ordered according to *idem*-identity, so that they both describe delineated and determined qualities.

Abstract space, then, is a *space conditioned on a notion of sameness premised on difference and identity*. Such space is inherently political as it institutes particular political imaginaries. State space carries its own spatial practices and representations of space that condition the ways in which political modes of existence are conceived and enacted. Alongside the process of homogenisation, state space fragmentises, compartmentalises, and quantifies various facets of life. These processes all correspond to each other by being organised through and around a notion of sameness premised on difference. Also the divisive element of space as fragmentation draws on sameness when it parcels out discrete entities of space while neglecting how embodied life cannot correspond to the strict separations upon which the compartmentalisation of space is constructed. Difference becomes manageable and remains within reach for the *logos* of the centre, as seen in colonial and neo-colonial spaces.

SPACE OF AUTHORITY

State space complements the meaning of territory in Agnew's territorial trap, since it helps understand how a space of authority organises life also beyond the reaches of territorial or legal boundaries. It also helps grasp contemporary predicaments of sovereignty. Arguing that sovereignty is to be understood as a symbolic form more than a political or legal fact, Bartelson suggests that "the main function of the symbolic form of sovereignty is not to represent objects more or less accurately, but rather to constitute these objects in rough conformity to its form" (Bartelson 2014, 16). This amounts to a governmentalisation of sovereignty, of which Bartelson sees an escalation since the end of the Cold War. He suggests the governmentalisation of sovereignty has served to efficiently cast the separation of inside and outside in modern political order along inverted lines. If, earlier, the peacefulness of domestic order was projected onto the presumed violence and atrocities of the international, this has, with the governmentalisation of sovereignty, been turned inside out. Consequently, attempts to protect populations from the violence of their regimes or to "rebuild failed states" are made on the assumption that threats to international order are found on the level of the domestic (ibid. 98). This means:

> Sovereignty no longer finds its ultimate justification in the provision of domestic peace and order, but rather in the promise of international peace and order. But since this promise makes sense only against the backdrop of a prior weakening or lack of sovereignty, international peace and order have been premised on the possibility of domestic disorder. This implies that everything that can be construed as a threat to domestic political order and social cohesion can be twisted into an argument for further interference and intervention in the name of the overarching objectives of international peace and order. As such, the governmentalization of sovereignty has amounted to a stealthy rewriting of the *nomos* of the Earth that has turned the modern international system into an empire in its own right, based on claims to boundless authority backed by appeals to universal moral standards. (ibid. 99)

Now, if the governmentalisation of sovereignty appears to have turned the world system inside out, then it has presumably also had effects on the bounding of space as a precondition and limit of sovereignty. While strategies of governmentalisation appear as limitless and as "backed by boundless authority claims" (ibid. 85), this does not signify the dissolution of territorial integrity. To Bartelson, the governmentalisation of sovereignty creates hierarchies amongst states depending on their capacity to adhere to the standards of governance and to conform to international law and ideals of democracy. This by no means relaxes the problem of political identity. If the democratic people are burdened by a democratic deficit in the lacking legitimacy of their delineation, then a global authority promoting democratic governance will also rely heavily on the attempted solutions to the legitimacy gap—namely that cultural differences are managed so as to maintain a certain degree of sociocultural homogeneity (ibid. 86–87). Global authority then depends on the portioning of the world into discrete national populations that operates "as a dispersed regime of governance of the larger human population" (Hindess 2000, 1494); though, while the boundaries of territory are fortified, the "sovereignty over the life within [them] has become negotiable" (Duffield 2006, 77; see also Brown 2010). To Bartelson, this means that "the political identities that underpin the international system are no longer primordially given or orchestrated by state authorities, but are instead the cumulative consequences of governmental efforts to maintain the preconditions of domestic democracy and international order. The ideological correlate of this is neo-nationalism" (Bartelson 2014, 87). Those resisting this would be the ones unrecognised, from paperless migrants, stateless, or indigenous peoples to pirates, rogue states, and terrorists. These have "replaced the barbarian, the savage and the uncivilized as the new Others of international law and order, making them legitimate targets of an unrestricted extra-legal use of force" (ibid. 89).

Hence, if there is a certain boundlessness to sovereignty as a symbolic form that at the same time fortifies the boundaries of the territorial state while homogenising its people, then this boundlessness appears to paint the space of authority as a space of calculability and sameness. In analogy with the simultaneous homogenisation and fragmentation at work in Lefebvre's abstract space, such space of authority makes local differences manageable by aligning them under its own logic. Because, for all their assertions to openness, the cosmopolitan claims underpinning the governmentalisation of sovereignty are also imbued with sameness, exclusions, and entrenched power relations and inequalities. Its limitless reach does not reduce the boundaries it sets to the meaning and possibilities of politics. Neither does it relax the borders it draws to whatever forms of stateless groups and persons unrecognised by the international system. Further, not only is the global call for states to conform to the standards of liberal democracy and international law generative of hierarchies and of what resembles an empire; it is also productive of sociocultural homogenisation on the level of domestic populations.

Pressing itself into practice, the sameness of the space of global governance joins hands with the sameness of the space of sociocultural nationalism. Despite the boundlessness in the governmentalisation of sovereignty, the reach of the latter and its infliction on state sovereignty appears by no means to loosen the boundaries of the territorial state but rather to consolidate them as the aggregate consequence of a global authority. Therefore, while the bounding of space once was the precondition and limit of sovereignty, it now seems that the precondition of sovereignty—both in its limitlessness as symbolic form and in its reflection in state sovereignty—is a space of sameness.

An attempt to eschew Agnew's territorial trap would therefore broaden the possibilities to answer the question of *where to find* the political community and the power and control it exercises. Yet, noticing how space of authority is productive of a homogenised yet hierarchised space globally as well as domestically, the question of the space of political community falls back into a question of *what*. What identity claims underpin global authority and what qualities are applicable to a particular sociocultural national character? Mirroring the transparency of territory, the homogeneity entailed in these identity claims forms a dismaying omittance of both the political agency of precarious beings who would form part of "the Others" of international society and the power, control, and violence to which they are subjected.

SPACE OUT OF BOUNDS?

So are there any prospects of a space out of bounds, a space that would seek its way out of territorialisations and homogenisations? An impressive attempt to reject efforts to "tame" space through boundedness, constancy, and separations is found in Doreen Massey's *For Space* (Massey 2005). Massey sees how bounded space relates to identity politics in providing a spatiality of homogeneity and cohesiveness that corresponds to the nation or a static collective identity. In contrast to this, she elaborates a notion of space as a product of interrelations. Such space emerges from interactions and manifests the "relational constructedness of things", including political subjectivities and political constituencies, whereby Massey negates identities as already constituted (ibid. 10). Drawing on Mouffe, Massey sees also spatial identities as constituted in relational terms. Space is understood as the "sphere of possibility of the existence of multiplicity" (ibid.). This means that space needs to be approached in the plural; as the space of a plurality of coexisting heterogeneous trajectories. Further, space is a process; it is always under construction and can never pose a closed system, but must rather be understood through its radical contemporaneity (ibid. 10–12).

Massey seeks to rework the opposition between a vast outside space and the intimate rootedness of place. She also rejects a distinction between time and space that gives priority to time while organising spatial heterogeneity along the logic of temporal sequence. Such space is found in the imagined unbounded space of globalisation that "chimes all too well with that powerful rhetoric of neo-liberalism around 'free trade'" (Massey 1999, 35). Instead, Massey draws on the intertwinement of space and time, which allows her to see the imbrication of space and place; while she understands space as "a simultaneity of stories-so-far", which form places as "collections of those stories, articulations within the wider power-geometries of space" (Massey 2005, 130). Space, as well as place, can therefore not be separated from time. This means that places are always different; they cannot be held still. To be in a place is to meet up with the histories of others at the point where these are at now, a point which includes the relations in the very "meeting up" (ibid. 125). Places, hence, are always shifting. They are different also for the ones who remain in place (ibid. 161–62).

Massey's very rich account of space, place, time, and their interplay with the multiplicity of cultural identity draws an image of a world of interaction, simultaneity, and connectedness. Her focus on the spatial and temporal malleability of identities is rewarding, especially in emphasising how relations precede identity. Still, it does not manage to avoid rendering identity in collective terms that rest on the assumption of both sameness and difference. Relying on the antagonistic approach of Laclau and Mouffe, Massey sees how new shared local identities are forged in opposition to a common antagonist. This is not simply an alliance between different groups or interests; rather the negotiations and relations modify "the very identity of the forces engaging in that alliance" (ibid. 182). However, although the identities have changed in relation to each other, the structure by which identities are established in opposition to an other remains. This means that connections of the topography of identity are relational in a particular way—by constituting oppositional relations. This indicates that there are relations that might escape Massey's relational topography, something that is also noticed by Ben Anderson, who questions the place of encounter, belonging, and 'non-relations' (Anderson 2008, 234).

Further, Massey's notion of a relational topography makes her emphasis on identity in relation to space peculiar. In the light of how she seeks to move away from spatiality as representative of identity, it makes sense that she aims to show the relational character of the spatial. However, while the latter move implies a withdrawal from the boundedness of space, identity persists in being predicated on boundaries. Individual and collective trajectories can meet and be forged into new ones, but their unfolding remains both linear and delineated until the next moment of interaction. It seems counterproductive to allow the multifaceted notion of space to hinge on a relatively cohesive common identity, no matter the volatility of the latter. With this, boundaries remain as "frontiers between subject positions" as a necessary requirement for politics (Lloyd 2005, 162). Thus, rather than posing a release from boundaries and separations, Massey's relational topography invites the difference and sameness copula through the back door opened up by the emphasis on identity.

What emerges here is a problem of the temporality of language and a particular understanding of contestation. Homi Bhabha attributes the problem of linear time to modernity. Modernity relies on a present that renders the " 'now' as transparent immediacy" and draws communality as "predicated on transcendent becoming" (Bhabha 2004, 241). Both performative and a pedagogy, modernity poses a self in constant reconstruction and reinvention in its striving to greater emancipation. The teleological temporality of modernity also implies spatiality, as development and progression rest on a distance to the event sparking its teleological movement. It is this distance that installs cultural homogeneity to the community, by introducing a narrative that grounds social and psychological identification and cultural communality. This lets the people assume the sign of the present, but a present established by a pedagogy originating in a point or event in the past, like, for example, the French Revolution, which let the people be reiterated and enacted according to the guiding narrative of said pedagogy (ibid. 145–46, 243).

Engaging the position of the "historically displaced", Bhabha identifies another aspect of modern time that sidesteps both the narrative of linear progression and the empty homogeneity of national time. Bhabha takes recourse to the heterogeneous space that lies between the signified and the signifier. Through signification, the meaning of a sign is determined, though never in an all-encompassing manner. There will always exist meaning outside the sentence, meaning that evades being grasped within a given sentence. Signification thereby forms a closure towards what would form the "non-sentence", although this is always "contiguous with the sentence" (ibid. 182).

As in Bartelson's discussion of identity and difference as the condition of (im)possibility of each other, Bhabha's sign harbours its own interruption since spatial and temporal extension disrupt the transparent now of modernity. There is a contingent moment between signification and the sign, which is embodied by the postcolonial subject. As signification and sign are not fully contemporaneous, there unfolds a space between them that invites the postcolonial subject centre stage. In the openness between a sign that exceeds meaning and a signification that performs a closure, Bhabha identifies the social activist who dodges the oppressors' attempts to determine the postcolonial subject. This subaltern agency is a resistance that demands new grounds for signification. Hence, the space opened up is a space of negotiation, reinscription, and resistance (ibid. 180–87). To Gayatri Spivak, this forms a "catachrestic space" where "words or concepts [are] wrested from their proper meaning" to form " 'a concept-metaphor without adequate referent' that perverts its embedded context" (Spivak 1990, 183).[1]

Still, the interstitial and volatile time and space opened up by Bhabha and Spivak must eventually lead up to a closure. The subversive subaltern agency requires a grounding, though not a "totalization of those grounds"; it "requires movement and manoeuvre" but not "a temporality of continuity or accumulation"; and it requires "direction and contingent closure but no teleology and holism" (ibid. 185).

The above suggests that while subversive subaltern agency is unpredictable and disruptive from the point of view of order and power, its manoeuvring and direction invests it with calculability from the perspective of the agent. Directed towards a closure or ground, the disruptive action in the catachrestic space also seems to follow a logic that both draws on the sameness of what counts as action and lands in an identity that gives rise to new externalities. The postcolonial subject as described by Bhabha negotiates her own space of existence, as well as her place in the community. In doing so, she disrupts the linear teleology and homogeneity of modern space. Such upsetting of the modern order should not merely be understood as a change of direction or redrawing of a land, but rather as an act that opens modern institutions to their contingent foundations. Still, the difference the postcolonial subject puts to work is a difference mediated by representational thinking. Resting within the language of the coloniser, the space between the sign and its meaning will lead to a new closure with new externalities. To follow Bartelson; if identity—the sign—already holds within it its differences, then difference—the space in between—harbours identity and with it closure and externalities.

For both Massey and Bhabha, the understanding of space as enacted, volatile, and multidimensional becomes tied to representation. Space forms a counterbalance and resistance to representation. It forms the means for contestation, augmentation, and transformation. Yet, tying space to representation, making representation the inevitable end of spatial enactment, sets forth a spatiality that forms a distorted reflection of a territory that manifests a particular collective identity. On the one hand, Bhabha and Massey divert the question of political community to a question of where to find the spaces of its enactment and unfolding. On the other hand, precisely as seen in the discussion on agonist politics, in order to constitute a *political* community, the place of a community rejecting imposed identities requires another space where identities are entrenched. Although place is needed for the political to be put to work and although politics is spatial to the extent that it takes place in space, politics continues to have representative thinking as its source. Due to being directed at the inequalities, hierarchies, and closures stemming from the operations of identity in a community conditioned on a notion of sameness premised on difference, the political must also reinstall identity in order to persist. Again the stateless, broadly understood, emerge as the protagonists of political action. But they also emerge as symptomatic of how politics becomes a question of both transgressing and redrawing boundaries of membership, of *logos*, and in space. Reinstalling identity means to also reinstall a notion of space as a container of said identity, with calculation, fragmentation, and hierarchies to follow.

Simon Springer points out how "anarchist geographies of co-operation are to be born from outsides the existing *order*, from sites that the state has failed to enclose" (Springer 2012, 1617). On that note, escaping authority and its imposition of sameness and difference on coexistence would amount to a geography detached from the assumption of identity as what spurs action and grounds community. The latter is what James C. Scott seeks in *The Art of Not Being Governed*. For the stateless people in Zomia—an immense mountain massif in Southeast Asia—and the non-state spaces they dwell in, identity is of importance only as long as it serves the purpose of not being governed. Of greater significance is the space of the hill areas of Zomia, where the various governments and states in the valleys long have had problems establishing and maintaining authority. Due to its inaccessibility, the space of the hills is prevented from being administered and organised under state governance. Rather than constituting spaces left behind, or yet to be touched by the processes of civilisation and state-making, Scott sees how the hills offered zones of refuge from life in the valley states. By relocating themselves in the hills, where they can live beyond the reach of the state, people have sought to evade projects of state-making in the valleys. Hence, "nomadism in the hills can be an escape or survival strategy" (Scott 2009, 22). Scott's study shows the inhabitants of the mountain lands in Zomia not as remainders of particular state-making projects but as "runaways" from such processes. Here, statelessness is a deliberate choice, as a question of being statefree, rather than being deprived of a state of one's own.

As part of the refuge from statehood, identities in the hills are marked by ethnic and linguistic fluidity, something needed in order to adapt to changing constellations of power: "The uniqueness, plurality, and fungibility of identities and social units in the hills are poor raw material for state-making", Scott concludes (ibid. 327). The fluid character of identity amongst some people also consists in the shifts from hierarchical to non-hierarchical social relations. On the contrary, since egalitarianism makes it difficult for a state to extend its power through local leaders, egalitarianism serves as a means to avoid state power. Scott further notices how the egalitarianism of some groups is accompanied by non- or only short-lineage histories, as a means to avoid status-building and hierarchies (ibid. 23–24, 275–79). In Scott's study, there seems to be an immediacy of identity, making the Zomians "linguistically and ethnically amphibious" (ibid. 23). However, his characterisation of the relation between hill people and valley people as well as his endorsement of how hill people must be understood relationally and positionally against the kingdom of the valleys can be read from two perspectives.

On the one hand, identity and ethnic variation emerge as variable and as marking the position of a group relative to state authority. Rather than a crucial means for action, identity thereby forms a way to avoid governance.[2] But if ethnic distinctions must be understood in relation to state authority (ibid. 32), then, on the other hand, Zomians are burdened with the same problem as seen above—of positing a closure of the meaning of politics and the possibilities for what it means to be political. A statefree Zomia, in that respect, is still ordered around the contestation and resistance to state authority.

Still, the non-state spaces of the hills, as described by Scott, appear to escape the elements of abstract space as theorised by Lefebvre. Zomia is not homogeneous but relies rather on its plurality and shifting character. It is also not fragmented, since the inaccessibility of the terrain precludes straightforward attempts of mapping and calculations.[3] As noticed by Gaston Gordillo, terrain also demands a more complex spatio-temporal understanding, since terrain changes with the seasons and different moments of the day (Gordillo, interviewed by Elden). Lastly, Zomia's non-state space avoids hierarchisations of both space and individuals. Contrary to Massey, who opens up space to grasp the malleability of identity, Scott shows how taking to the hills as an attempt to avoid governance renders identity important only as long as it serves this aim. The non-mapped space of the hills appears to be a perfect reflection of such attempts.

Yet, while terrain removes space from geographical and calculable space, it remains within a spatiality in which the near and remote are attached to physical distance. Similar to how the identity of hill people is relative to state authority, then, a mountainous spatiality of statefree Zomia will hinge on a state spatiality marking out proximity and distance, that is, a spatiality that sets apart rather than brings together.

CONCLUSION

Even if Benhabib recognises the problematic "Gordian knot linking territoriality, representation, and democratic voice" (Benhabib 2004, 219), she makes few attempts to resolve this. Taking stateless communities into account reveals how this problematic cluster lets membership be interlaced with ownership. Living one's life, even for generations, in a particular place would not be a sufficient criterion for engaging in local iterations as part of the *demos*.

Instead, democratic voice and citizenship appear to pair up with recognised entitlement to land. This chapter suggested that in order to escape Agnew's territorial trap, it is not enough to try to release political thinking of the community from an idea of state space as dehistoricised territory that merely contains social and political relations rather than being instrumental for how they are produced.

In addition, in order to escape state space, political thinking of community must escape a space of sameness, that is, a space ordered according to a rationale of calculability. Such a rationale will seek to make sameness and difference manageable under the same *logos* of calculative thinking. To release the question of political community from a thinking that compartmentalises the world and orders political coexistence according to a notion of territorial belonging that simultaneously homogenises, fragmentises, and installs hierarchies, a *relocation* of the question of political community is required. Relocating the political community implies not only an attempt to seek it *elsewhere*. It also means to locate it anew—to find another way of grasping the place of the political community that is not dependent on a spatiality that sees space as measurable, bounded, and continuous.

Both Massey's and Bhabha's elaborations on spatiality attempt such a relocation of the question of political community. On the one hand, with a spatial engagement that shuns away from territoriality, they engage a notion of political action not primarily delimited by membership and boundaries. On the contrary, Massey draws a spatiality where encounter is prior to identity and Bhabha paints a place of politics where meaning is still open and abundant. For both, the political is not located *within* a closed area but unfolds rather in a space that exceeds meaning. Still, it is only a temporal and partial relocation of political community that Massey and Bhabha pursue, since their notions of spatial acts are explicitly directed to the transformation of meaning, representation, and identity. While they might tear open the exclusive ties between population, territory, and democratic engagement, they do not avoid locating the political community in a place conditioned on membership and boundaries. Thereby they end up re-inscribing sameness and difference. They must also answer to the same problem as the agonist school in the previous chapter—the question of who is at liberty to appear in space and to take place.

Likewise, the use of a phenomenology of terrain, with its appearance that shifts with perspective and with external impacts such as weather and daylight, helps Scott sustain a political community for which identity forms neither a source nor an end but a tool for contestation and resistance. Yet, the place of the political community is held peculiarly intact. It is located in the valleys or the hills; it is ordered through a notion of proximity and distance that sees space as extension rather than as intertwined embodied existence.

In that sense, it clings onto a space of sameness that makes difference appear manageable from the rationale of the centre and thereby contains and isolates their exceptionality within a broader framework for how to grasp political order and belonging. Peculiarly, then, the anarchist communities described by Scott form mere aberrations to the state instead of opening up the meaning of politics and the political community to new constellations of coexistence.

Also these attempts to spatially engage the question of political community, then, ultimately land in a question of *where* and at *what moments* more inclusive forms of community can be identified, rather than revolving around *how* such communities affect the very ways we think about belonging, politics, and space in the first place.

For Arendt, statelessness means to be deprived of "a place in the world which makes opinions significant and action effective" (Arendt 1976, 296). These persons are deprived of more than fundamental human rights, for they have lost the right to opinion and the right to action. There is no correspondence between how they are acted upon and what they do. Whatever is done to them is without "any relation whatsoever to what they do, did, or may do" (ibid.). If to deny a person a place in the world means to deny her the possibility to act herself as a political person whether through speech or gesture, this implies that we need to understand the meaning of *place*. To Arendt, place implies a space of appearance, a space where beings emerge in the world through their actions. This is a place of natality, that opens up the existing orders and introduces novelty into the world. Stemming from action, it is marked with unpredictability and boundlessness as it forms a manifestation of freedom (Arendt 1998, 177–78).

In many ways, Arendt's notion of place and of a space of appearance departs from a state rationale that orders according to calculability and certainty. The public aspect of political action is nevertheless crucial for Arendt. *Polis* is not first of all physically located but implies a space emerging between people acting and speaking together. It forms a world in which words and deeds are meaningful. The space of appearance forms a spatiality of togetherness and community but relies also on a phenomenology of visibility (ibid. 198–200). This makes sense, considering how Arendt seeks to characterise the mechanisms of dehumanisation put to work by totalitarianism.

Being deprived of a place in the world, the stateless is also deprived of a world of meaningful human relationships. While this is crucial to Arendt's attempt to picture the continued devalorisation of human life that she saw unfolding before her eyes, it makes it difficult to open the question of political community towards the *statefree*. For Arendt, action gains precedence to space; space emerges through meaningful human relationships in which action takes place, redirects, and sets the political community off to new journeys. The way we think about action will determine the way we grasp our place in the world and the space of community. Action opens the world to incalculability and inauguration. But unfolding on the register of the public, Arendt's notion of action and consequently of space and the world will draw its own externalities, precisely because of her attempt to picture the predicament and these externalities and to grasp the calamity of their situation.

But what if we approach the question of political community the other way around and ask how the ways we think about place and the spatiality of the political community affect the ways we think about belonging, coexistence, and political action? This would, perhaps, imply engaging the unpredictability and incalculability of *place itself*.

What follows seeks to elaborate a notion of space that withdraws from the calculability and sameness upon which the space of authority and the space of the nation seem predicated. When seeking to escape abstract space, Lefebvre put his hope in resistance in difference[4] and suggested especially the faculties that exceed the reductive qualities of *logos* to be conducive for this. Poetry and music stand out as practices that escape the cultivation of knowledge pursued by *logos*. Instead of pressing knowledge into the service of power, poetry, music, and art form aspects of the "Anti-Logos" that develop "multifarious distortions and discrepancies" in space (Lefebvre 1991, 392). To poetry is also where this book turns, when attending to the space and place conjured by the poetic river in Heidegger's work on Hölderlin's poem "Der Ister". Through cautiously threading Heidegger's paths, the next chapters engage a notion of space that withdraws from the calculability and sameness upon which the space of authority and the space of the nation seem predicated.

NOTES

1. This sheds light on Spivak's, to some, controversial discussion on strategic essentialism that argues for a strategic adoption of essentials that serves to promote a deconstructive awareness of the structures essentialisms produce and exist in. Such deconstructive awareness serves not only to reveal the work performed by master significations and categorisation, but also to show the futility of their claims to totality (Spivak 1990, 45–104). Spivak emphasises that no political theory can be built on the concept of strategic essentialism and that one indeed must give up any claims to theoretical purity as soon as a universal is deployed in struggle. Still, her use of the concept rests on her theoretical position that assumes how, in order to be able to act at all, one has to, at least momentarily, adopt a universal (ibid. 12, 45).

2. Scott highlights how illiteracy amongst people in the hills could be understood as a way to avoid societal structure. He notices how "virtually all hill peoples have legends claiming they once had writing and either lost it or that it was stolen from them" (Scott 2009, 23). It is at least conceivable, Scott argues, that this is a deliberate

adaptation to statelessness, due to the plasticity oral histories exhibit in comparison to written genealogies (ibid.)

3. A practice of mapping attentive to the specifics of the mountain massif would need to take into account distance as walking distance, Scott suggests. Simply depicting a kilometre as a kilometre seen from above and uncorrected from the "friction of terrain" makes little sense in a rugged environment of mountains, marshes, and swamps. Instead, nearness and remoteness would need to be considered in relation to the length of actual travels (Scott 2009, 47–48).

4. Indeed, he elaborates a *right to difference* as a right to not "be classified forcibly into categories which have been determined by the necessarily homogenising powers" (Lefebvre, quoted in Elden 2004, 229–30).

Part II

INCALCULABLE COMMUNITY

Chapter 4

Incalculable Space

The first part of this book saw attempts to expand the community end up in forever new externalities. The thinking of political community and its possibilities was seen to hinge on identity. Thereby, the way the political community is imagined tied into the way politics is conceived but also how the political subject and political action are constituted. Vested in visibility and circumscribed by representational thinking, political existence was seen to be conditioned on a certain degree of liberty, which might indeed preclude the most vulnerable in society from the realm of political action as well as the sphere of moral concern. Locked within calculability and representational thinking, the question of political community emerged as a question of boundaries. These boundaries were manifested in the assumption of semblance between community members, which made even the selfhood of the individual emerge as an *idem*-identity, posing a particular citizen character as a precondition for the collective. They also emerged as the boundaries of a contestable, augmentable *logos* that require delimitations in order for the possibility of contestation and of politics to be upheld. Finally, they were drawn as boundaries in space, separating a physical inside from an outside. But they were also seen as boundaries *of* space, of how the space of the political community is conceived as a measurable, bounded room that forms a correlative to the bounded *demos*.

In the third chapter, it was argued that space forms an active component of power, control, and categorisation precisely due to it being grasped as calculable and as based in sameness productive of differentiation and fragmentation. This chapter takes upon the task of thinking an incalculable space by engaging Heidegger's work on spatiality. Thus, rather than turning directly to texts dwelling on the limits and possibilities of reason in order to find hints of the meaning of an incalculable community, this chapter seeks to approach the question of the political community through attending to the spatiality of the community. It will be argued that the spatiality emerging in Heidegger's work precludes a notion of space as delineated or homogeneous. Drawing on Heidegger's notion of dwelling, it is maintained that one's existence in the world is inherently spatial and ephemeral and evades being inscribed in representational identity. The argument is foremost developed with respect to the two texts *Hölderlin's Hymn "The Ister"* and "Building Dwelling Thinking" and concludes that as gathering place, also the formally non-included must be thought of as part of the community by being part to its place.

As highlighted in the introduction, turning to Heidegger and especially his reading of *The Ister* to ask the question of a political community not predicated on sameness, difference, and identity is a precarious move. Following Heidegger's path, it was maintained, must be done carefully so as to avoid the pitfalls of his thinking. A few aspects of the notion of space and place emerging below will be of extra consideration. This chapter seeks to elucidate the spatiality of the community as well as the spatiality of beings. For the purpose of the continued analysis, it is of special importance whether notions of space and place emerge as bounded and owned. It is crucial to understand if there is an exclusivity of place that maintains a delineation between us and them. To better grasp the latter, Heidegger's elaborations on the relation between the foreign and the homely is explored. The final parts of the chapter discuss how the notion of place developed over the course of this chapter can be grasped as being for all, or if it is tainted with continued exclusions. They also discuss the implications of the notions of space and place for understanding beings as already part to the community, their formal status notwithstanding.

However, Heidegger himself has criticised the 'metaphysical principle of identity' in a way that might be productive for engaging notions of sameness and difference that escape representational thinking. Therefore, the chapter first turns to Heidegger's own work on identity, difference, and sameness, while also studying how spatiality plays an important part in Heidegger's attempt to move away from a Cartesian subject and metaphysical identity.

IDENTITY, DIFFERENCE, SAMENESS

Scholars engaged in topography and critical geography suggest that the spatial gains in importance and centrality in Heidegger's later work (Casey 1997, 243–45; Elden 2001, 13, 16). In his early works, including *Being and Time,* Heidegger alludes to questions of space and spatiality, but temporality and historicity are generally understood as the key feature of *Dasein* (Elden 2001, 13, 16). Although the word *Dasein*, which both is a noun meaning existence and a verb meaning "to be there", has spatial implications, Heidegger was still clear in *Being and Time* that spatiality is not a primary characteristic of *Dasein*; instead "an understanding of being-in-space is only possibly after an understanding of being-in-the world as an essential structure of Dasein" (ibid. 16). Still, the existential analytics in *Being and Time* is developed from a critique against Descartes's ontology, a critique based in the spatial existence of *Dasein*. However, before shedding light on this, it might be fruitful to skip some forty years ahead in Heidegger's *oeuvre*, to see what he himself has to say about the matter of identity, difference, and sameness.

For Heidegger, identity is ultimately a question of the relation between being and beings (*das Sein und das Seiende*), a relation that forms the pivot of most of Heidegger's work. *Being and Time*, for example, wrestles with how the question of being is forgotten and neglected within Western metaphysics from Plato and Aristotle onward. Indeed, it is not only that being is forgotten in Western metaphysics, but the very forgetting of being is forgotten as well. Both Plato and Aristotle sought to make sense of the question of being through taking beings or a particular being as their point of departure for reflections on being. Being is thereby reduced to beings, leaving the question of the meaning of being untouched and concealed. To Heidegger, the relation between being and beings rests on a particular concept of time. The condition of possibility for being to be grasped through beings is a concept of time understood as presence, or as 'now'. Western thought forwards a metaphysics of presence that culminates in the meeting between the 'now' and Descartes's notion of cogito as absolute presence for itself. Together, these pose only what is present as truly being. This, however, serves to reduce the complexities of existence and makes being identical to beings. To avoid this, the primacy of the now must be disrupted, Heidegger maintains. The relation between being and time must be rendered more complex, and an investigation of how various forms of being correspond to various temporalities needs to be pursued. From the different horizons of various forms of being and their times, that is, the being of beings, an original horizon could possibly be uncovered. In *Being and Time*, then, the task Heidegger sets before himself, and ultimately fails in completing in the unfinished work, is to grasp an original horizon and an original time (Birnbaum and Wallenstein 1999, 14–17; Heidegger 2010, 1–3).

An original time would allow for a fundamental ontology, according to Heidegger, who distinguishes between *ontic* and *ontological knowledge*. While ontic knowledge is the knowledge of the particular nature of beings and of sciences, ontological knowledge is the *a priori* conditions of possibility for ontic knowledge. That is, ontological knowledge is the knowledge on the basis of which ontic knowledge is constructed. A *fundamental ontology*, that is, the question of being, would describe the conditions of possibility not only of ontic knowledge but also of the ontologies that precede and found the ontic sciences (Heidegger 2010, 10).

The question of the relation between man and being takes centre stage also in *Identity and Difference*, a text allegedly considered by Heidegger as his most important publication since *Being and Time*. Now, however, the analysis starts not with the meaning of man and works its way towards being. Instead, it centres on the relation of man and being *as a relation* and inquires hence into the relation as a relation rather than into its components (Stambaugh 1969, 7–8).

Identity and Difference turns to the principle of identity in Western European thinking that says, "To every being as such there belongs identity, the unity with itself" (Heidegger 1969, 26). Whereas A = A posits two As as equal to each other, though not necessarily the same, the principle of identity says that A *is* A; that "every A is itself the same" (ibid. 24). The principle of identity thereby states that the unity of A "forms a basic characteristic in the Being of beings" in Western thought (ibid.). Indeed, it is through this metaphysical principle of identity that science, as we know it, is possible since "if science could not be sure in advance of its object in each case, it could not be what it is" (ibid. 26). This is not to say that science must know in advance what it is looking for. Rather, the metaphysical principle suggests that to the extent that ontic science is concerned with the identity of the objects studied, then it must also rest on the ontological assumption of the very possibility of such an identity. The metaphysical principle of identity draws together both being and the human, or being and beings, and makes identity a fundamental characteristic of both. Since both being and beings are grounded in their identity, there is also an identity between being and beings, according to Heidegger. Thereby, any connection between the human and being is always established "either in terms of Being or in terms of man", making one of them the representation of the other (ibid. 32).

Heidegger juxtaposes the metaphysical notion of identity with a fragment from the pre-metaphysical thinker Parmenides, which states that being, together with thinking, belongs in the Same. What can this Same to which being and thinking belong consist in? It cannot be an identity in the metaphysical sense, since metaphysical identity strives towards the absence of difference. In contrast, the Same which emerges here is a Same to which *both* being *and* thinking belong and therefore must accommodate difference. The Same, Heidegger suggests, might be better thought of as a *belonging together* of being and thinking. With the metaphysical concept of identity, the emphasis rests on belonging *together* so that belonging is determined on the basis of together and the manifold is turned to unity. When instead the *belonging* in belonging together is emphasised, as Heidegger suggests it should be, then being and thinking are both held apart and at the same time brought together (Heidegger 1969, 23–30; Stambaugh 1969, 12–13).

To think the belonging of human and being requires a move out of representational thinking, a move that takes the form of a leap "in the sense of a spring [*Sprung*]" (Heidegger 1969, 32). The spring is a leap away from being and the interpretation of it as a ground in Western thought. Still, this does not mean that the spring away from the ground is a spring into an abyss. Or, rather, it is only an abyss from within the "horizon of metaphysical thinking" (ibid.). Instead, "insofar as we spring and let go", the spring is a spring into being. A spring is then what is needed to "experience authentically the *belonging* together of man and Being" (ibid. 33). The realm in which the spring leaps is the realm "from which man and Being have already reached each other in their active nature,[1] since both are mutually appropriated, extended as a gift, one to the other" (ibid. 30–33, at 33).

Heidegger writes of the spring that it is a curious one; a spring that gives us the insight that we do not really reside, or at least not sufficiently, where we are (ibid. 33). The key to grasp this is found in appropriation, that is, in the "event of appropriation [*Er-eignis*]" and its relation to the "framework [*Ge-stell*]".[2] It is in the event of appropriation that Heidegger locates the belonging of the human and Being. The event of appropriation operates in between ontological and ontic knowledge. Although the event is phrased as a noun, it should be understood in a verbal or processual sense. It is an *Er-eignung*; a happening or essencing of ground. The event is thereby not ontic, but the "grounding dimension of [the] condition of possibility" of ontic occurrences (Marchart 2007, 21). With the event, being gives itself in different forms that themselves are both invoked by and inductive of particular ways of thinking. In modernity, where Heidegger situates himself, the event of appropriation stands therefore in close relation to the "framework" or the "frame [Gestell]". The relation between the event and the framework decides how being is delivered in man and man in being. In this age of technology, the frame determines the appropriation through planning and calculation, so that beings "appear within the horizon of what is calculable" (Heidegger 1969, 35). Beings appear then as present, delineated, and open to categorisation. In "a more original appropriating", on the other hand, where "the dominance of the frame" might be relaxed, the event of appropriation "is that realm, vibrating within itself, through which man and being reach each other in their nature, achieve their active nature by losing those qualities with which metaphysics has endowed them" (ibid. 37). Hence, Western metaphysics poses beings according to a particular understanding that lets them be grasped in terms of representational identity. From the perspective of ontic science, beings would ideally correspond completely to their representations. From the perspective of a being standing before the power of a categorising knowledge, one could instead see the difference between a being and its representation as a space for resistance, as exemplified by postcolonial thinkers such as Bhabha and Spivak. Something similar could be said of agonist thought, in which "working the interval" would amount to augmenting and relocating categories of inclusion. This would, however, not do away with the sameness/difference copula as an ontological condition of possibility of resistance, change, agonism, and political community. To pose the question of political community in a way that lends thought for an incalculable community, it would instead, following Heidegger, need to grasp a notion of sameness in which difference forms a part, not an aberration or constitutive externality to a unity vested in identity.

For Heidegger, identity is but one possible form in which being is given over to the world through the event of appropriation. "What does appropriation have to do with identity?", Heidegger asks and gives the resolute answer: "Nothing" (ibid. 38). In contrast, "identity, on the other hand, has much, perhaps everything, to do with appropriation" (ibid.). With metaphysics, identity is presented in and through the framework as a statement that appropriates being and beings in accordance with this statement. But with this, any reflection on the world is reduced to that statement and thinking is stopped halfway (ibid. 38–40). The reason for this consists in how the event not only gives being but also makes being withdraw. Emerging in one guise, other appearances of being withdraw. In the "framework"—in an age of technology that embraces identity—it is not only that other appearances of being withdraw, but also the *very possibility* of other appearances is forgotten. This means that also the event withdraws; the *happening* of the emergence of a particular historical character of being (such as *physis, logos, idea*, substantiality, objectivity, subjectivity, the Will, or the Will to Power) withdraws, meaning that the horizonal character, that is, the finitude, of each giving of being is concealed. To release thought from the oppression of the appropriative frame a thinking of "Being with respect to its difference from beings" is therefore required. This means that being cannot be fully thought; it will always differ from the beings trying to grasp it. Various epochs of being open finite horizons of thought within which beings are grasped in certain ways. While horizons do not exhaust the possibilities of beings and of being, their finitude leaves the very openness of being—or being as openness—unthought.

Thus, assuming unity as the condition of possibility of the political community has already framed the community in particular ways. It establishes the identity of the community as corresponding directly to the identity of its subjects. Often, as Rancière underscored, the political dimension of this is effaced from the self-apprehension of the community. This further bounds the possibilities of coexistence in particular ways so that they are in conformity with the overall idea of the community. Instead of a holistic approach seeing the whole as more than the sum of its parts, a community presumed to be already known is posited. Wendy Brown's analysis of the neoliberal political world in which each subject is required to order and present itself according to a neoliberal rationale forms an excellent example of this (Brown 2015).

When addressing sameness, difference, and identity, Heidegger says that what is *same* is the event of appropriation. All appearances of being and beings emanate from the event of appropriation, which both gives an historical being and lets this be appropriated in the frame. This means not that the event of appropriation is given, or that it is. The *Ereignis* is a giving; a clearing that determines how being gives itself, but with that very giving both the event and being withdraw (Birnbaum and Wallenstein 1999, 117; Heidegger 1969, 40). *Difference*, then, is what the same in the form of the event of appropriation harbours. Difference emanates from the inescapable difference between being and beings. With each horizon of thought opened through the event of appropriation, others will withdraw: there is always more to being than the epochal characters it assumes. *Identity* is one way that being is appropriated into the world. Peculiar to the metaphysical principle of identity is how it neutralises the event and forces the difference between being and beings into oblivion. Thereby, identity consumes the meaning of the same and the different and makes them graspable only according to its own appropriative frame in which they are constitutive of each other. The metaphysical principle of identity turns the same and the different into subcategories of identity. Hence, it leaves no room to think sameness and difference outside of representational thinking. Sameness becomes the identical or equal and difference the non-identical, instead of a same that warrants for the reoccurrence of difference by being formulated on the basis of belonging; as a gathering of difference.

DESCARTES AND METAPHYSICAL IDENTITY

Heidegger sees a paradigmatic example of the metaphysical principle of identity in Descartes's use of the notion of substance. To Descartes, beings are determined by the two substances: *res cogitans* and *res corporea. Res cogitans* is the substance of the thinking ego, whereas *res corporea* is understood as extension in terms of length, breadth, and depth. Extension is attributed to any corporeal thing so that all corporeal forms can be grasped in space as various modes of extension. Further, extension is not only a component of spatiality but identical to it, meaning that the world and intra-worldly beings are derived from the same substance, namely *res extensa*. There are a few reasons why Heidegger finds this problematic. First, Heidegger observes that substantiality is left unexplained by Descartes, who considers its meaning inaccessible. As a consequence, any attempt to ontologically ground substantiality is evaded, leaving the ontological foundations of the world as *res extensa* declared inexplicable. This is untenable to Heidegger, who argues that Descartes conflates ontological and ontic knowledge into indistinctness. Substance functions both in an ontic and an ontological sense so that it encompasses the knowledge that can be gained *as well as* the premises on which that knowledge is structured. Hence, substance exemplifies how science sets limits to its objects of study before even commencing its work; to study the world means to study the corporeal things grasped through extension, but this study is premised on the assumption that knowledge consists in nothing but a knowing that can accommodate extension—a knowing found in mathematics and physics (Heidegger 2010).

This leads to Heidegger's second problem with Descartes: the centrality of mathematics and calculability. Heidegger objects to mathematics as a measure to know the world. By mathematical knowledge, a "secure possession" of the being of beings can be obtained (ibid. 93). In this, mathematical knowledge understands being as something that *is* and is constant in being what it is. Hence, the being understood mathematically is a being that "is *what always is what is*" (ibid. 94). Not only does mathematics pose being and beings as constant and enduring, but the way mathematical knowledge understands beings is also *constitutive* of these beings (ibid.). Beings must be posed so that they are graspable by mathematical knowledge. As a consequence, Heidegger maintains, Descartes does not allow the "being of innerworldly beings to present itself"; rather, his idea of being is posed as constant objective presence and precludes any primordial understanding of *Dasein* (ibid.).

It is in relation to this that Heidegger's notion of space takes form. Against a mathematical measure of space, he poses an experiential understanding of the spatial. *Being and Time* engages the experience of equipment, or useful things, to grasp a space that defies mathematical rationality. Our relation to equipment, Heidegger says, is not primarily determined through the measurement of distances but rather through our everyday experience of them. Notions such as closeness or nearness (*Nähe*) or orientation (*Ausrichtung*) are more helpful to understand a lived and experienced space. Useful things and equipment have their place in the space where we dwell, and by having their place, they in some sense give rise to that space as, for example, familiar. Thus, a space is not simply "a three-dimensional multiplicity of possible positions"; such dimensionality would still be "veiled in the spatiality of what is at hand" (ibid. 10–101). Rather, what is above, below, here, or over there is "interpreted on the paths and ways of everyday dealings", of which knowledge can never be gained through "observational measurement of space" (ibid. 101). As being-in-the-world, *Dasein* is de-distancing, every being encountered by *Dasein* is encountered in a nearness that makes distance disappear, that brings near. Hence, nearness and remoteness has nothing to do with measurable distances. For the calculative knowledge pertaining to the latter, two points can never be near or remote from each other; they can only be separated by a distance. However, to Heidegger, the position of *Dasein* in the world cannot be determined in terms of objective presence. Rather, the place *Dasein* occupies is such that it de-distances what is at hand and gathers this in its surrounding region. As a de-distancing being-in-the-world, *Dasein* has directionality; it is directed towards beings. What determines the spatiality of *Dasein* is hence being-in, de-distancing, and directionality (ibid. 102–7).

De-distancing and directionality disqualifies the mathematical notion of space, since the latter must assume a homogeneity that an experienced space never can exhibit. Heidegger indeed asserts that to only appreciate homogeneous space will mean the loss of the worldliness of beings encountered. In contrast, the attempt to understand space ontologically means that space must be freed from the "narrowness" of what is available at hand, as well as from any "undifferentiated concepts of being" (ibid. 109). Already in *Being and Time*, then, is it clear that Heidegger's understanding of space is intimately connected to the question of being; in order to grasp space ontologically, one must also turn to being and to the world as "space co-constitutes the world in accordance with the essential spatiality of Dasein itself with regard to its fundamental constitution of being-in-the-world" (ibid.). It is also clear that Heidegger pictures a space *opposed* to a calculable space. This indicates a space not imbued with a state rationality that seeks to order and delineate in accordance with representational identity or particular categories. However, while the smoothness of the abstract space of the modern capitalist state appears interrupted, or at least as what is to be questioned, it is less certain that Heidegger's space escapes lending itself to the symbolic sameness of a nationalist space. To grasp that, a more in-depth reading of Heidegger's understanding of space and place is needed.

Alejandro Vallega argues that despite the importance of the spatial for the critique of the Cartesian substance, spatiality itself withdraws from the analysis in *Being and Time*. With the priority given to temporality, Heidegger installs a break between spatiality and being, since *Dasein* is given a purely temporal horizon. As a consequence, Heidegger leaves "no room for conceiving facticity and its spatiality otherwise than in terms either of the necessities of ontic (objective) presence, or pure temporality" (Vallega 2003, 142–43). Space thereby "withdraws from the withdrawal" operative in the event of appropriation, leaving spatiality "open to be thought" (ibid. 12–27, 141–43, at 143). Heidegger's later work, by contrast, show attempts to think space in itself (Elden 2001; Malpas 2006; Vallega 2003). As seen before, thinking involves setting out on unfamiliar waters and differs therefore from *res cogitans*, which departs from what one *knows*. For Heidegger, art and poetry form openings for a thinking of a notion of space freed from the constraints of mathematics. The hymnal poetry of Friedrich Hölderlin is also, as Elden notices, where Heidegger abandons the priority of time over space and aims to understand them together, something especially done in the Ister hymn (Elden 1999, 267). Hence, it is to the waters of the unknown the remains of this chapter turn.

RIVER SPACE

Yet what that one does, that river,
No one knows.
(Hölderlin, "The Ister")

The rivers Hölderlin poetises should not be read in a symbolic way, Heidegger says. They are not images or metaphors intended to convey a metaphysical meaning (Heidegger 1996b, 18). No doubt, the rivers sung of in Hölderlin's hymns—the rivers Main, Neckar, Donau, and Rhine—could easily lend themselves as symbols of German life and essence. Yet Heidegger turns away from an understanding of art that would, in accordance with Plato, place it in a suprasensuous or nonsensuous realm. To engage the river poetry through how it can be understood metaphysically would be to neglect what the river in itself can convey. Instead, it puts poetry to the risk of being drawn into the pitfalls of the metaphysics through which it has been interpreted. In contrast, Heidegger sees Hölderlin's rivers to stand fully outside the realm of metaphysics and are not "symbols of a higher level or of 'deeper', 'religious' content" (ibid.).

When turning to the Ister, Heidegger sets out to *attend to*[3] the essence of the river. To attend to the essence of the river means for Heidegger to be attuned to the activity of the river, or rather *the river as activity*. Hölderlin's poetry reveals how this activity remains concealed. The river has its own time, and perhaps its own place, which might be intimated though never fully presented by the condensed thinking poetry allows for. The river is an enigma with an origin and activity that can lend itself to poetic thinking, yet, *what it does, that river*—that is, what is decided by the river's activity—this *no one knows*. In its distinction, in its particular doing, the river is concealed also to the poet, who cannot grasp what is decided by its activity. Instead, the poet must return to the activity itself, to the lived experience of the river to maybe find a path towards what is decided in the river's activity (Elden 2001, 35; Heidegger 1996b, 19–20).

What is it, then, the activity of Hölderlin's Ister? In his poem, Hölderlin evokes the activity of the river (*der Strom*) to be its flow. The river flows, *er strömt* (Heidegger 1996b, 20). To see the river's activity as flowing means to render it with both a temporal and a spatial dimension by which both a "now" and a "here" is named. Indeed, "The Ister" begins with a "now", a moment (*Augenblick*)—"Now come, fire! [*Jetzt komme, Feuer!*]"—but continues soon with a "here"—"Here, however, we wish to build [*Hier aber wollen wir bauen*]". Whereas other parts of Hölderlin's poetry touches on the "now", "The Ister" gives an "equal intonation" to the "here" (Elden 1999, 266). This indicates not a reversed priority between the temporal and the spatial. Rather, what Heidegger seeks in his reading of "The Ister" is a way to understand how time and space belong together in an originary unity (ibid. 266–67).

Hölderlin poetises the river as vanishing yet full of intimation. In their flowing, rivers "abandon the Now" as they pass "into what is bygone, or into what lies in the future" (Heidegger 1996b, 15). The river before us will always already have moved onwards, so that the present has already passed away, or, for that matter, is about to be filled with what is coming. Vanishing and intimation draw the rivers as "oriented in a twofold direction", since "as vanishing the river is underway into what has been. As full of intimation, it proceeds into what is coming" (ibid. 29). Still, the meaning of both intimation and vanishing is not unequivocal. Intimation suggests something of what is to come; it hints towards the future and is so directed to what is coming, but it relates at the same time to what has been. To intimate proper, one must search in the depths of one's own experience, meaning that one must engage in inner recollection (*Erinnerung*). Recollection means not simply to turn to what is past, as this would touch merely an "outer surface" of past events. Proper inner recollection is instead that which turns inwards (*das Inwendige*) to what has been and to what remains *undisclosed* in what has passed by. It means to open up the past and allow it into the present and the future towards which the present extends. Hence, "genuine inner recollection is intimation", Heidegger writes (ibid.).

As vanishing, the river vanishes into what has been; one now moves into another now that has already flown by so the river almost appears to move backwards (ibid. 15, 27–29). Flowing into what has been, the present river seems not to find its future in the sea towards which it flows, but rather in the creeks left behind. The burgeoning creeks, which on the journey of the river seems almost forgotten, would thus form an origin or a source, but a source that temporally lies ahead and is to come. Still, "vanishing can also be an inconspicuous passing away into what is coming, a decisive belonging to whatever is coming" (ibid. 29–30). For something to come, for something to emerge as heterogeneous to a uniform time, vanishing is indispensable. To the river—which finds its only constancy in a continued transformation, in its flow—vanishing forms the possibility for a coming of a not-yet-known, be that a to-come in the future or gathered from a reencounter with an undisclosed past. If something new or other is to come, then the past must be left behind. Nationalist temporality shows how holding too tightly to a past encountered by, for example, nostalgic longing will cause this imagined past to put its mark on the present. The same goes for the future; a too clear projection of the future will already have ordered what is coming in line with the future projected. If deviating from the linearity of this time, actions or events are seen as aberrations to the linear temporality rather than as *contestations to its assumed normality*. In contrast, the river describes a past that must pass by, for only by letting go can what is coming arrive.

The flow of the river unfolds not only in time, as the river takes its course also in space and names a "here". Indeed, the river names a particular here when Hölderlin writes, "Here, however, we wish to build [*Hier aber wollen wir bauen*]". The "here" that is named is a locale, a place, an *Ort*. But this locale, named by Hölderlin's Ister, is nothing but the human abode, the dwelling place of human beings. Dwelling place (*Wohnstatt*) and abode (*Aufenthalt*) should not simply be understood as house of residence or accommodation. The German word *Aufenthalt* rather suggests the abode to take on the meaning of resting place. Heidegger connects it to *verweilen* (linger, tarry, to while) and underlines that "the abode is a whiling. It needs a while. In such a while, human beings find rest" (ibid. 20–21). He also emphasises the inviolability of such resting sites in Greek thought, while pointing out how Hölderlin speaks of these as "asylums". An abode has a locale that determines the abode; that gives it a certain locality (*Ortschaft*). In his reading of "The Ister", Heidegger finds the locality of the human abode to be determined by the river; "the river 'is' the locality that pervades the abode of human beings upon the earth, determines them to where they belong and where they are homely [*heimisch*]" (ibid. 21).

While the meaning of the homely is the focus of the last parts of this chapter, I here wish to dwell upon the place conjured by the river. Heidegger emphasises that the essence of dwelling, of the human abode, can be approached through the essence of rivers. Thus, the dwelling place, like the river, is vested with time and space, though not in any uniform, linear, or geometrical sense. The river does not simply provide for a good spot to build a house; rather, it renders human dwelling as locality. Locality, however, "is attained in and through journeying" (ibid. 31). Just as the journey (*Wanderung*) of the river is marked by its intimation and vanishing, that space yielded by the river also appears as twofold:

> The essence of the locale, in which becoming homely finds its point of departure and its point of entry, is such that it journeys. The essence of this journeying is

> the river. The locale is both there and here, not according to some contingency, but under the concealed law of a journey. The locale is not something that is first there and then here, in a mere succession or arbitrary arrangement of places that are occupied and then relinquished. *The previous locale remains preserved in the subsequent one. And the subsequent locale has already determined the previous one.* This is why the 'there' and the 'here', indeed the transition between the two locales, are everywhere named by rivers. (ibid. 35–36, my emphasis)

With the activity of the river, the present and what comes shift the appearance of the past. Not only is a previous locale preserved in what comes, but the locale of the present and the locales to come will augment what has preceded them. Space emerges here as far from static. Instead, it continuously changes, both in the present and in the past, along with the flowing river. Heidegger asserts not only how the past is in the present but also how the present augments the past, though not in the sense of a nostalgic projection of the past that seeks to shape it in accordance with the present. On the contrary, the past place is as much as the present place conjured by the river. Whatever disclosure of the present place will shed new light also on parts of its past, still concealed, character.

Together, vanishing and intimation reveal how the activity of the river forms a *journeying*. It describes the journey that thought must undertake through what is other or foreign to it in order to come to insights on itself. This suggests a present, which is in emergence and directed towards the future, while its emergence is grasped through a careful attention to and encounter with a still unconcealed past. As vanishing, the "now" of the river appears to be anyplace but *there*; the river flows away into what has been yet while doing so it brings another "here" so that it indeed appears to vanish into what is coming. As full of intimation, the river's "now" seems to spread along its entire course. It extends towards both to the sea that is coming and the creeks of its emergence, by grounding any forbearance of the future in a gathering of a concealed, undisclosed past.

Yet Heidegger's paths need to be threaded with care. Would a future found in an intimation of the past lend thought to a constant search for the origin and trajectory of the 'we'? Would this trajectory then bind 'us' to the place gathered by the river, so that blood becomes intertwined with soil? The persistent emphasis on the German in *Hölderlin's Hymn "The Ister"* would perhaps contest that such conclusions could be avoided. Peculiarly, however, it is precisely with the notion of *place* these concerns can be dismissed. Heidegger writes of how the "intimative vanishing" of rivers "along their own path" forms an abandonment of the present and rests on an "unfaithfulness" towards the landscapes through which they run (ibid. 28). Further, "thoughtful remembrance proper" belongs to "this enigmatic unfaithfulness" (ibid.). If the historical grounding inherent to the journey of the river is made possible by an *unfaithfulness* to the landscape through which it takes its course, it appears that the place gathered by the river can never be vested in the uniformity that the notion of communal soil would invoke. The place, as well as the present, must be abandoned for the coming to unfold.

Or, put differently, the place itself is unfolding; it is vested with a peculiar, passing spatiality that makes place emerge as both spatial and temporal. Unfaithful to itself, it will continue to emerge and vanish as the river passes by. Heidegger writes, "What is proper to the river is the fact that it flows and thus continually determines another 'Here'" (ibid. 15). This would not only mean a "here" understood in the sense of a point that moves along the course of the river as the river flows towards the sea. More importantly, the river determines the locale it flows by, meaning that this locale is always shifting. Its lack of constancy suggests that the place named by the river is always named anew. Indeed, the river *is* its own locality, Heidegger writes, as Hölderlin says of the Ister, "Beautiful he dwells" (ibid. 35).

Here, by the river that dwells in its flowing, a different meaning of space, place, and the *Ort* takes form. To Heidegger, the river's locality and journeying form "a concealed and originary unity" (ibid. 39). An originary unity is opposed to a calculable geometrical grid, described by coordinates *x, y, z* to which time can be added as a fourth factor *t*. On the latter view, space and time are perceived in terms of identity, through being beforehand posited in a way that makes them graspable according to mathematical coordinates. Seen as objective and calculable, space and time function as containers accommodating all various spatiotemporal appearances. A calculable, coordinated spatiotemporality becomes a tool for conquering and controlling a particular world and a particular notion of human existence in this world (ibid. 39–45).

Calculability perceives space and time through togetherness and as similarity rather than as *belonging* together. Therefore, the possibilities of space and time will elude us. To understand space and time and their belonging together, one must thus depart from a modern thinking that seeks to organise the world in terms of order. With "order", Heidegger means "calculable and ordered rationality. Whatever is subject to order must be posited advance in such a way, and can be posited only in such a way, that it becomes accessible for such order and can be controlled by it" (ibid. 40). The spatialisation of the planet signifies an "absolute domination of space in the form of total and readily available presence" (de Beistegui 1998, 84). This draws presence not as something in part undisclosed or coming into being. Rather the "essential unfolding of presence has withdrawn" (ibid.).[4]

The human abode—the place localised by the river—forms a venue for thinking space and time together. Approaching this abode in terms of a rootedness or autochthony that would run the risk of linking blood to soil would draw a linear history in which sequences add on to each other with space as a backdrop ridden with symbolic meaning yet in itself relatively static. This would, as seen, be in conflict with the temporality of the locale of the river. A place is not succeeded by another place like beads on a thread (ibid. 35). On the contrary, not only will the past affect the present, but what happens in the present sets its mark on the past, puts it in a new light, and requires of us to approach it anew. Further, the river shows the present as never really presentable and at hand. As the river flows, the present turns into a presencing, a continuous unfolding and vanishing, which cannot be brought to a halt without abandoning the very essence of the river. If the river names the human abode by rendering it as a locale, invested with the space of locality intertwined with the time of journeying, then also the abode—the communal place—must be incessantly emerging and passing. Like the river dwells in its flowing, the place of the 'we' appears to find its permanence only in a transient unfolding.

Heidegger writes of how the river "is" the locality pervading the human abode, a locality that brings humans "into their own" and "determines them to where they belong and where they are homely" (ibid. 21). Now, if the river's activity is its flow and if the locality of the place is conjured by the river, then it seems that the locality of the river must be vested with flow. This locality, which, along with its emergence, also vanishes, is described as what determines humans both in their essence and to where they belong. It seems that the flow of the locale rubs off on the ones finding their abode therein, so that their own, their very homeliness, becomes saturated by a flow that escapes an ordering rationale.

Thus, to attend to the activity of the river means to step out of the calculability in which the modern notion of space is fettered and instead embrace an incalculable space. Such space breaks any attempts to bind it both temporally and spatially. It cannot be territorially determined or organised linearly. Instead, incalculable space unfolds; it gives itself with the vanishing river; it withdraws once intimated. But the river not only conjures an incalculable place that will unfold along with the river's flow, it also draws human existence as marked by the changes of that very flow. Turning to another river, the one encountered by Hermann Hesse's Siddhartha, one can see how it is not only that Siddhartha cannot step into the same river twice, but that it is also not the same Siddhartha that twice steps into the river (Hesse 1951). Thus, the river disqualifies nostalgia as a yearning for lost places (Casey 1993, 37) by showing how the nostalgic being herself is not the same. Places longed for are inescapably lost since there is no same self that could return to them as they were; they can only be encountered as new.

DWELLING SPACE

With the Ister, there emerges an image of a *place* of the community that escapes the boundedness of calculable space. Vested with the flow of the river, this is an ephemeral place that both unfolds and vanishes. But how does this place escape excluding boundaries? And how do beings relate to place?

Although attempting to give a new meaning to an old language, Heidegger's vocabulary is saturated with connotations that lead thought to the opposite of an open and inclusive place; the human is characterised as the homely and her spatiality unfolds around the notion of dwelling (*wohnen*). In what follows I will argue that the conclusion on a *Blut and Boden*-theme in Heidegger is, if not unfounded, at least not a necessary outcome of Heidegger's discussions on space. What will be seen is instead how dwelling casts light on the human relation to space and how the human activities of building and thinking are involved in a gathering of place. Also this, I suggest, brings forth the location as shifting; as unfolding with each human engagement. Thus, the notion of dwelling proposes that each human interaction in space will augment that space, shift its appearance(s), and rewrite its time(s).

It is in the short text "Building Dwelling Thinking" (Heidegger 2001a), originally written with the attempt to address the post-war housing shortage, that Heidegger identifies dwelling (*wohnen*) as the basic character of the human being. Here, the human emerges as a spatial being alongside her temporal existence. The human relates to a locale; she takes place. Dwelling is "inherent in the concept of the human" (Elden 2001, 86), which indicates that space is not outside or above humans but emanates from the human and how she relates to the places in which she emerges. Building, then, follows upon dwelling, not the other way around; "we do not dwell because we have built, but we build and have built because we dwell, that is, because we are dwellers" (Heidegger 2001a, 146). So what does it mean to dwell? The meaning of the old Gothic word *wunian* is not simply "to remain" or "to stay in place", Heidegger asserts; it also addresses the actual *experience* of remaining. *Wunian* signifies "to be at peace", "to be brought to peace", or "to remain in peace". Heidegger accentuates the link between "peace" (*Friede*) and "free" and sees how peace indicates to be "preserved from harm and danger", "preserved from something", or "safeguarded" (ibid. 147). Hence "to free really means to spare" (ibid.). Sparing (*schonen*), in turn, encompasses a tenor broader than to spare someone from harm and danger;

> Real sparing … takes place when we leave something beforehand in its own nature, when we return it specifically to its being, when we 'free' it in the real sense of the word into a preserve of peace. To dwell, to be set at peace, means

> to remain at peace within the free, the preserve, the free sphere that safeguards each thing in its nature. (ibid.)

This underscores that no one or nothing is to be mastered or subjugated. Being spared means to be free to form one's own existence in the world. To spare "means to set something free into its own presencing [*etwas in sein eigenes Wesen freilassen*]" (ibid. 148). An image is here drawn in which dwelling emerges as an openness, by allowing beings to exist in the way they are, and sparing forms the mode by which each individual forms her life at her own discretion.

We attain to dwelling through building, Heidegger suggests (ibid. 143). What, then, does it mean to build? Through building, space is located and organised. In the Ister lecture, Heidegger says that although the locality is intrinsic to the rivers, they are themselves not part of the landscape (Heidegger 1996a, 33). Instead, the landscape appears to be brought forth first with building, that is, with a human intervention in that locale which is granted by the river. In "Building Dwelling Thinking", the river is no longer at the centre for thinking space; rather, it is a human construct that serves to clear a space for thought on the spatial—the bridge.

The bridge that stretches over the river does not simply connect two already present riverbanks with each other. Quite the contrary, the bridge is the construction that lets the banks emerge as banks and situates them across from each other on each side of the stream. The bridge brings the two banks to the stream and to one another, while also allowing the spaces beyond them to extend in the distance. Thus, "the bridge gathers the earth as landscape around the stream" (Heidegger 2001a, 150). Building forms a process that gathers space as a location, a site, in which the bridge allows the room to emerge and be vested with meaning (Birnbaum and Wallenstein 1999; Heidegger 2001a, 150–51).

The spatiality presented by Heidegger's bridge is very different from any territory firmly secured by borders:

> The bridge lets the stream run its course and at the same time grants their way to mortals so that they may come and go from shore to shore. Bridges lead in many ways. The city bridge leads from the precinct of the castle to the cathedral square; the river bridge near the country town brings wagons and horse teams to the surrounding villages. The old stone bridge's humble brook crossing gives to the harvest wagon its passage from the fields into the village and carries the lumber cart from the field part to the road. The highway bridge is tied into the network of long-distance traffic, paced as calculated for maximum yield. Always and ever differently the bridge escorts the lingering and hastening ways of men to and fro, so that they may get to other banks and in the end, as mortals, to the other side. (Heidegger 2001a, 150)

Instead of delineating space through an enclosing border, the bridge gathers, joins, and allows for movement. The bridge locates an experienced place from which space unfolds. Thus, rather than departing from a space to identify places within it, Heidegger sees how places make room for a site, for a space. Space thereby comes after the location, after the human engagement with the locality granted by the river (ibid. 152). Spaces—now articulated in the plural by Heidegger—are gathered through locations and the buildings of human dwelling.

There is an ambiguity to the spatiality of dwelling concerning whether the notion of place eludes boundedness or not. At one point, "Building Dwelling Thinking" appears to articulate space as something that *interrupts* the continuous spread of land gathered by the bridge. A space is namely "something that has been made room for" through the location, "something that is cleared and free, namely within a boundary, Greek *peras*. A boundary is not that at which something stops but, as the Greeks recognized, the boundary is that from which something *begins its presencing*" (ibid.). Here, a boundary emerges, as space is "in essence that for which room has been made, that which is let into its bounds" (ibid.). Rather than extending and unfolding in the surroundings of the gathering point, spaces now seem to appear as separated in different sites, compelled by their own bounds, which are giving room for their own separations and differentiations.

These few paragraphs of "Building Dwelling Thinking" have received a lot of attention and are usually interpreted as providing a view of human existence as placed and bounded (Malpas 2006; Rose 2012; Young 2002). Although alluding to space in terms of openness, multiplicity, and fluidity throughout both "Building Dwelling Thinking" and in the Ister lecture, Heidegger here appears to paint space as delineated, sealed off, and something that can be claimed as one's own. In political geography, two conflicting readings of dwelling stand out. In one, the concept is interpreted exactly as providing for a view of space as open and multifarious (Ingold 1995), whereas the other reading addresses the sedentarist implications of dwelling and argues that dwelling establishes human existence in terms of rootedness. This reconfirms the belonging to a particular place and, in extension, paves the ground for territorial nationalism (Sheller and Urry 2006, 209).

One of the possibilities to avoid a closure of Heidegger's account of dwelling place is due to, as Elden points out, the fact that Heidegger himself articulates space as *Raum* rather than space. Contrary to space, *Raum* does not relate to the Latin *spatium*, which signifies an interval or intervening space. Rather, *Raum* describes space as extension; *extension* and location are thus, according to Elden, potentially unbounded (Elden 2001, 85–88; Heidegger 2001a, 152–53). However, Heidegger also underscores that extension as well as *spatium* can be described by abstract algebraic relations, which paints space as measurable according to distances and directions. Hence, although the emphasis on *Raum* allows for a shift in focus from the delineated *spatium* to the more abundant extension, it offers little help to unfetter the meaning of space from the trap of calculability. Instead, I attend to how the boundary itself in Heidegger's text precludes a delineated or bordered location, or dwelling as sedentarist and framed through ownership.

Certainly, Heidegger's characterisation of the boundary risks leading thought toward the understanding that the dwelling place is exactly that freed, cleared space for which room has been made within a border. Still, Heidegger's text gives way for another notion of the boundary than a strict delineation. Drawing on Greek philosophy, Heidegger accentuates that "[a] boundary is not that at which something stops but … the boundary is that from which something *begins its presencing*" (Heidegger 2001a, 152). This is why the concept of the boundary "is that of *horismos*, that is, the horizon, the boundary" (ibid.). In addition to the boundary as a category of emergence and unfolding, I would add that the horizon always *withdraws*; it retreats in the distance. Move closer to the horizon and it will move away from you. In this sense, the horizon constructs a boundary that both allows for the emergence of space and maintains it as constantly indistinct. The space, the site, lacks a given starting point as much as it cannot be presented as finite. While this reading draws boundaries as permeable and elusive, it does not exhaust the meaning of *peras* or *horismos*. To do so, a return is needed to the question of the event of appropriation and of the implications of the emergence and withdrawal entailed in this.

THE EVENT OF PLACE

The difference between calculable and what I here call incalculable space rests on the distinction between two ontologically different views of space. Cartesian mathematical space occurs, with Heidegger, on the level of the ontic. Although modern sciences would assume themselves to forward an ontological understanding of calculable space, this only proves the conflation of the ontic with the ontological to Heidegger. Instead, calculable or ontic space is merely one way of grasping space, which proceeds from a calculable science that strives to make space and science comprehensible by the same economy. In short, such science strives to grasp space as present, as at hand. For Heidegger, this has nothing to do with any actual meaning of space but works rather to perceive the world in accordance with the *Gestell*—the framework of calculability. Following Heidegger, a spatiality that strives towards incalculability must instead draw on a spatiality not at hand as present, but wrought by the same gathering and presencing through which beings are disclosed. Such gathering was seen in "Building Dwelling Thinking" where dwelling implies a *schonen*, a sparing, through which beings are disclosed or set free into their own presencing. On the one hand, presencing draws beings in their finitude, in their coming to pass. What they are will not remain; disclosed in their essence beings are bound to an ephemeral existence. As dis-closed, beings must avoid being referred back to the closure inherent to representational identity. On the other, it brings forth beings in their not-yet. As presencing, beings are yet to come. They are imbued with a futurity emanating from their finitude. They do not hold an infinite existence on the grounds of an eternal idea or in an identity that would inscribe them as present-at-hand. Such modes of existence would encumber their essential temporality.

The presencing and absencing of space as well as of human existence is entwined with Heidegger's notion of truth as *aletheia*, which is a central theme of *Parmenides*. Heidegger's discussion especially centres on the relation between *aletheia* and *lethe*, between truth and oblivion, unconcealment (*Unverborgenheit*) and concealment (*Verborgenheit*), or disclosure and withdrawal. While *lethe* is the counter-essence of *aletheia*, the two do not stand in direct opposition. Instead, truth "is un-truth insofar as there belongs to it the reservoir of the not-yet-uncovered, the un-uncovered, in the sense of concealment" (Heidegger 2001b, 58). The disclosure inherent to *aletheia* does not mean to draw something out of oblivion, of *lethe*, so that it becomes unclosed. Instead, *lethe* is an operative element of *aletheia*; it forms the concealment and withdrawal that gives the possibility of truth. If *aletheia* would have referred to the unclosed, to that which is brought into daylight and which thereby is seen and known, it would not be a matter of truth at all, but one of certitude. The latter is what calculable reason engages when posing questions on an axis between falsity and *veritas*. From this axis, the very element of *lethe* is withdrawn. Certitude has therefore nothing to do with truth as *aletheia*. *Aletheia* could be seen as what is put to work in the mimicking, the interplay of words, and the staging seen in the work of Bhabha and agonist thinking. Such actions and enactments emanate from the insight of the possibilities of existence covered up in the uniformity of posited identities, signs, or categories.

To Heidegger, the disclosure of *aletheia* is an opening of the open (*das Offene*). This is where certitude fails, as assuming the being of beings according to the parameters of one's own *rationale* would be the least but open. Instead, the unconcealed must be accompanied by the concealed, to let openness persist. With presencing comes therefore absencing. There, in the play between emergence and withdrawal, the truth of being finds its abode. This is why "the dis-closure [Ent-*bergen*] is at the same time an en-closure [*Ent*-bergen]" that shelters the unconcealed in unconcealedness (Heidegger

1992, 133). Heidegger alludes to the secrets of existence, to what remains hidden in the appearances of beings, and to which we can attend to through wondrous curiosity regarding what is left open.

Thus the open leaves matters open to questioning and renders a situation open-ended, as it were. To Heidegger, the open lies in the free, though not in the sense of a limitless vastness or a scattered existence. Contrary, the free of the open is what constitutes that sheltering enclosure. What is sheltered is nothing but being itself; "the free is the guarantee, the sheltering place, for the Being of beings" (ibid.). In the open, as the free, being is sheltered and salvaged. It is so, through the openness of *aletheia*, which is dependent on *lethe*. If truth withdraws when giving itself, then truth can be preserved from certitude and remain open to questioning. In analogy, the free is the sheltering place (*die bergende Stätte*) because it salvages being as both presencing and absencing, as a lingering unfolding, so that "the open ... is Being itself" (ibid. 150).

To Jeff Malpas, the gathering of place must be understood as an event. In other words, the event is an event of place.[5] As happening, place is an event that brings the world by drawing the dwelling of mortals into its midst. The event reveals the world in its concealment and opens up the world to questionability and disclosure. Yet this is possible, Malpas argues, precisely due to its bounded character. To sustain this argument, he highlights both Heidegger's characterisation of space as something that "has been made room for", that is "cleared and free, namely within a boundary" (Heidegger 2001a, 152), and the notion of the horizon as that from which something begins its presencing. He also emphasises how this attributes a more positive and enabling meaning to the boundary. In analogy with the limit-character of death as that which allows life to unfold, Malpas sees how a thing or place does not simply come to a halt at the limit or the boundary; rather, the limit is what allows for the place to appear (ibid. 298). To appear, as a thing or as a human, is to be related to the boundary. However, this means not simply to come into presence from a limit, but that existence means to be within something bounded and demarcated. Hence, "death is the limit that opens up a 'space' within which our lives can be lived" (ibid. 273). If death is a limit, and as such the horizon that we bear with us, then our lives unfold from our hesitant knowledge of that limit. This unfolding appears as a place of its own; "having a sense of one's life as a whole is thus also having a sense of the perspectival and situated character of one's life—of the *topos* that belongs to that life" (Malpas 2012, 189).[6]

Hence, Malpas's event of place posits beings as situated and as immersed in the place in and with which they appear. However, there is a peculiar emphasis on place as bounded that seems to neglect the withdrawing aspect of the limit formed by the event. If place indeed forms an event, then place must also be touched by the concealing *Enteignis* which the *Ereignis* also is. To Malpas, "appearing will be such that it occurs within a certain domain that is open so as to allow for that appearing, and yet also bounded so that appearing is indeed an appearing of some thing" (ibid. 46). Further, "to be is *to be in place*, and to be a phenomenon, an appearing, is similarly to be placed, or, one might say, to *take place*" (ibid.), though with this, the focus is centred on appearance, rather than on taking place. Instead, focusing on the " 'taking place' of place, on the event of opening the open" would constitute the very issue of approaching space from *Ereignis* (Raffoul 2012b). This would mean to attend to the withdrawing character of the event and to engage *horismos* as enabling of *both* presencing and absencing.

When Heidegger's rendering of place as "made room for", "cleared and freed … within a boundary" and as "that which is let into its bounds" (Heidegger 2001a, 152) is read together with the discussion on *aletheia*, this implies that a cleared and freed place can be so only through its openness. The freeing aspect of making room for a place consists in leaving place open to wonder and awe; in attending to the *Verborgenheit*, the concealed, of any appearance. Though for this, the absencing boundary is central. While the *horismos* enables presencing, it is its withdrawal that makes possible such an enabling. In a double move, the *horismos* brings forth what might come, while its receding saturates the world with secrets of the not yet thought and not yet encountered. Like the river, the *horismos* vanishes into what is coming; though if place is an event, then place *itself* must be imbued with the secretive and concealed.

Thus, place is made room for within a boundary which is permeable and fleeting and forms the very opening of openness. To the extent that the disclosure is a salvaging enclosure in the free, the boundary as *horismos* must always function to withhold from place any closure. Hence, place is bounded by the *horismos* or *peras* only as far as it is kept in openness, but this precludes places as demarcated, bordered, or static. On the contrary, both the event and the figure of *horismos* conjure ephemeral places from which equally transient spaces unfold. These are places of finitude, though not in the sense of being delineated against an outside, but in the sense of not resting on an ontology of sameness. It is the lack of sameness and calculability to the places gathered that grants them their finitude and figures them as passing. Heidegger's characterisation of time as jointed must hence apply also to space. The forgetting of the event of appropriation, which results in the notions of linear time and measurable space, consists in the persistence of the *while*. Between emergence and withdrawal there lies a while. The order of this while, its *dike*, consists in it being a jointure between a coming into being and a passing away. The while is hence enjoined toward a twofold absence that forms both a binding and an unbinding, or at least an invitation of the latter. When standing in *adikia*, in dis-order or injustice, the while has left its transitory state and "becomes concerned with the permanence of its continued existence" (Heidegger 2002b, 268). This would be the concern of the calculable space and reason of modernity. There, the continuance of what is present finds its grounding in the boundlessness of a while oblivious of its order. It stands in *adikia*, concerned with its beginning but negligent of the passing of the while and hence of its jointed existence towards its withdrawal. Its time is out-of-joint, as Derrida puts it.

If the epochal giving of being renders being a transient character, it still appears to draw the various epochal givings of being as relatively stable. While *dike* serves to open thought to the event of being so as to address the out-of-jointedness of linear time, it gives fewer clues of what follows the potential passing away of such time. Is an epochal, historical giving of being merely succeeded by another giving, or can a continued transience of the while itself be intimated? Neither the *Ereignis* nor the *horismos*, as here characterised, seem to move thinking in such a direction. So does, however, the other figure of presencing and absencing touched upon here—the river. Again, the activity of the river is found in its flow, which at each moment forms a presencing and an absencing. Every 'now' and every 'here' is both a vanishing into what has been and an intimation of what is coming. It was said of the abode named by the river that it "is a whiling. It needs a while. In such a while, human beings find rest" (Heidegger 1996b, 20).

The rest they find is not such that life comes to a halt, but consists in the steadfastness (*Beständigkeit*) of human essence (ibid.). This could be interpreted as the while necessary for the abode is such that it slows down the flow, gives humans the steadfastness they need to handle the horizonal character of life and be, perhaps even grounded, in the epochal giving of the place where they appear. Though if the flow of the river names the locale, it seems that steadfastness rather consists in the dwelling by which the human relates to the emergence and withdrawal of herself and of the world. The steadfastness of human essence rests not in a static representation. It rests not in an eternal idea. It rests in both a disclosure and a withdrawal, so that the human emerges in and through a not-yet and a to-come that leaves herself open. The while of the abode, then, must not only be enjoined towards both presencing and absencing, but indeed immersed in an unceasing emergence and withdrawal. This order of the while paints the human abode as *in passing* so that the abode itself forms a continued presencing and absencing of both its site and its 'we'.

EXCLUSIVE PLACE?

This openness of place, as well as the figures of *horismos* and of flow, counter some of the reasoning and conclusions drawn regarding the rootedness and exclusivity of dwelling. While Mitch Rose's understanding of dwelling gives it a political meaning, Julian Young discusses dwelling as an ethical orientation to the world. However, in both cases, dwelling connects to an 'us' or an 'our' that marks a recurrence of delineations and potential exclusions. However, the above elaborations undermine these conclusions.

Rose's reading of dwelling centres on how Heidegger means that we *attain* to dwelling through building. To attain (*gelangen*) to dwelling by building, Rose argues, suggests that dwelling is something to be achieved, accomplished, or obtained. Dwelling constitutes a *marking*, since "dwelling 'marks out' a specific region of relations through the production of objects that catalyse those relations" (Rose 2012, 759). The landscape unfolding from building, Rose contends, is not only *sensed* but also a *seen* spacing; it "is marked out—that is, apportioned and set apart" (ibid.). Mortals exist by virtue of being "in and among a set of distinctive positions"; by being "among other beings who are properly spaced" (ibid.). Further, with marking comes a *claiming*, since Heidegger's world will always withdraw from any marks put upon it. While marking demarcates and names specific regions, claiming shows the vulnerability inherent to such marking. Any world built is never *Dasein*'s own to dominate. Instead, *Dasein* can only respond to a world of constant temporal unfolding, by making a claim to it. Rose rightly recognises that dwelling is a never accomplished task. This means that marking will not provide for stability and permanence. It also means that the world must continually be reclaimed. In the "unrelenting movement" of the world, "Dasein builds in order to make a claim" so as to disclose "not simply *a* world but *our* world—a place that names our being vis-à-vis a restless unfolding" (ibid. 768–69, at 769). Since the unfolding world is never really ours, this is a futile task, and marking and claiming shed light on Heidegger's characterisation of dwelling as "a process of coming to be at home in not being at home" (Heidegger quoted in Rose 2012, 769).

Now, the problem is that claiming addresses the way 'we' relate to a world of ontological unfolding, while marking—in apportioning, setting apart, and determining distinctive positions—comprehends this relation in ontic terms. One could say that building as marking gives to an ontological homelessness an ontic home. Claiming then serves to remind 'us' that this is merely a translation, that any marking might lose its meaning and return 'us' to homelessness. Hence, 'we' can find 'ourselves' at home, but 'we' are then simultaneously transferred to an ontic level. 'We' might be astutely aware of the fragility of 'our' position, however this adds an uneasiness to the claim made. If the claim is both to a world and to an 'us', and if it is articulated in ontic terms by building, then claiming appears to form a mark of its own. If this is correct, then claiming also forms an apportioning and setting apart, not only of a landscape but also of us. Reflected in an apportioned space, 'we' emerge again as inescapably delineated against whatever is outside of the lines marked out. Politics as a disagreement and a contestation concerning the meaning of us and our place is here reintroduced, though at the expense of thinking place and the spatiality of *Dasein* in itself.

Young, in contrast, sets out to uncover the ethical underpinnings of the concept of dwelling. In an eloquent attempt to grasp an ecological ethics from Heidegger's notion of dwelling, Young reads dwelling as a concern primary with what is close to oneself—both in distance and in relation. Despite also being a concept that holds ethical implications that reaches beyond the immediate community in which one resides, the main concern of the ethics emanating from dwelling is local. With this, Young appears to confirm the sedentary view of dwelling in which dwelling is paired with a rootedness in a particular, bounded place, and in which dwelling becomes a practice enacted primarily in relation to a people and a community that one can identify with (Young 2002, 70–71, 100–102). Young distinguishes between essential dwelling and existential dwelling. As seen, Heidegger draws the human, in essence, as a dweller. Yet the plight of modernity consists in man being homeless precisely by the loss of the knowledge of how to dwell in nearness with being; the human has forgotten her ability to actually experience her dwelling and herself as a dwelling being. Hence, *essentially*, the human is a dweller in the sense that every human possesses the capacity to dwell. *Existentially*, however, modern humanity is homeless in the sense that it has failed in a dwelling "which consists in *understanding* one's essential dwelling and living in the light of that understanding" (ibid. 7–74, at 74). So why does dwelling—this essential trait of humanity—become caught in rootedness when read as existential dwelling, that is, as an ethical concern that the human can become attentive to?

By translating the notion of *schonen* as caring-for, Young opens for a relation to beings as imbued with identification.[7] Indeed, Young reminds of how "what you identify with is what you care about" (ibid. 70). This must not be one's family or country but could also be the world or the world-centre. With this, Young points at the possibilities of self-transcendence but highlights also how caring-for is emotionally charged and runs the risk of being primarily concerned with what is appreciated as near one's own (ibid. 71). This becomes more problematic when Young turns to the *ethos* of the dwelling place.

In modernity, the human has lost her gods. Yet the gods are not dead; they are merely absent; they have withdrawn into concealment. What remains of the gods is an *ethos*, which Young sees belonging to particular dwelling places. Contrary to my reading, Young finds the dwelling space, designated by *Raum*, or place, to be strictly delineated (ibid. 100). For Young, dwelling place, which is "an interconnected complex of natural and cultural features which adds up to the notion of *place*", is within the boundary, a space made room for, cleared and freed for settlement (ibid. 101). A dwelling place is a place with its own, albeit absent, gods. The dwelling place still upholds the authoritative *ethos* of its people. For Young this forms an opening for a community—"to belong to a place is to be committed to its *ethos*" (ibid.)—through which a notion of belonging mediated by sameness and difference resurfaces. "Different places … house at least marginally different 'gods'", Young continues while stressing that one cannot feel equally at home everywhere as this would require a multiplicity of possibly incompatible commitments which "threatens the unity of a person" (ibid.).

Apart from the problematic premise of the possibility of such a unity, there is another predicament to Young's reasoning, namely the lack of attention to Heidegger's elaborations on space in texts such as "Building Dwelling Thinking" and *The Ister*.[8] Indeed, Young makes a peculiar connection between one's homeliness and Cartesian calculated space. He emphasises the dwelling place as extending in concentric circles so that one first and foremost is homely in one's own home, then to a lesser degree in one's village, even less so in the province and the country where one resides, and finally one is "not at all home out of [that country]" (ibid. 101n113).

The volatile space, which has emerged in the reading of *The Ister*, is here captured in the framework of delineation and geometry from which Heidegger sought to release it: a framework that draws its energies from modernity, that is, the time of man's homelessness and destitution, which also is the predicament that Young's ethical reading of dwelling attempts to dissociate from. Instead, for Young, space, time, and *ethos* appears inherent to a particular place rather than to the people constituting it. He mentions that emigration encompasses and requires a change in personality—even if just a nuance—of those migrating. This is most certainly so, though Young's conclusion is that "changing places is changing people" (ibid. 102), without pondering the possibility of it also being the other way around; changing people is changing places.

For Young, the spatial implications in Heidegger's concept of dwelling and building mostly pertain to architecture, which might be why he focuses on *techne* as a central tenor of dwelling as caring-for. Young highlights *techne* as an aided bringing forth, as when an artist releases a sculpture from a block of marble. However, Heidegger underlines how building is not solely a question of *techne*, which, he says, designates a "letting-appear"; a bringing something already made into what is already at hand (Heidegger 2001a, 157). This would characterise a process through which an idea is given a physical form in a particular way and is inserted into an already existing world. In this process, the spatial character of that world appears to remain largely intact. However, building is never solely letting-appear, but its nature consists in *letting dwell*. Building entails bringing forth locations, that is, sites marked by flow, by *strömen*, so that they harbour both temporal and spatial dimensions in a continuous unfolding. Such an unfolding is aided in terms of it being dependent on human activity, though it also relies on a stepping back, of *sparing* the unfolding from outside interventions, of setting that emergent space free into its own presencing.

For Heidegger, the nature of building rests in the *joining* of the spaces of the locations emanating from building. This, I suggest, is the meaning of building as letting dwell; it both brings forth and brings together chiasmic spaces that know no neat boundaries and cannot be sealed off from each other. If *techne*, "letting appear", relies on a certain augmentation of what is to appear for it to take place amongst what is present, then the ethics of "letting dwell" rather suggest that what is emerging does so by being spared from such an augmentation. Yet it emerges through a joining of spaces, through the altering of the locations of its own, as well as of what resides in its nearness. The ethics of "letting dwell" also points at how existential dwelling is not a practice one can learn and then always possess. Young draws an image of the dweller as relatively stable, whole, and homely when underscoring how "the dweller's primary care is to relieve others of that primary 'misery'" (Young 2002, 123). This, however, contradicts Heidegger's account of the central predicament of dwelling. "The real plight of dwelling," Heidegger writes, lies in "that mortals ever search anew for the nature of dwelling, that they *must ever learn to dwell*" (Heidegger 2001a, 159).

LEARNING TO DWELL

Homelessness forms a modern human condition to Heidegger and is also the dominant theme to the Hölderlin lectures. In the Ister lecture, the discussion on learning to dwell in a condition of homelessness unfolds with respect to the relation between the homely and the foreign, or the self and the other, through which the self must take its journey in order for it to come closer to its own essence. Now, it probably appears peculiar that an emphatic necessity of the homely and the foreign should have any bearing on the discussion of an incalculable community. Have these two figures not already established relations of inside and outside as well as sameness and difference, one might duly wonder? It is to *Antigone* the Ister lecture turns when discussing the homelessness and the relation between the self and the other. It is also with reference to *Antigone* that the lecture formulates learning to dwell as a questioning disposition to existence.

Heidegger describes how the foreign "is not some arbitrary 'foreign' in the sense of whatever is merely and indeterminately not one's own" (Heidegger 1996b, 54). On the contrary, the foreign is intimately related to one's own; with the foreign starts the return home. Indeed, the foreign is the necessary origin for the commencement (*der Anfang*) of becoming homely. For a return to be made, the foreign must be separated and distinct from one's own. Only by being "known and acknowledged in its essential oppositional character" can the foreign be intimately related to one's own. Such a uniting would not be a "confused mixing but a conjoining in distinction" (ibid.). Here an image arises of peoples, strictly separated from each other, though yet fulfilling *their own* destiny through one another; that is, an image in which people remain bounded, grouped together as people set apart from each other and who necessarily must remain so in order to pose the possibility of disclosing their self by the means of their other.[9]

Heidegger's notion of translation helps grasp the meaning of "conjoining in distinction" and the relation between the foreign and the homely. Translation, according to Heidegger, never functions to simply substitute a word in one language for an equivalent word in another. Translation must instead be an interpreting, drawing on the connections between two languages but also within the same language. By interpreting the meaning of a word in one language into a second, the understanding of the first language is deepened and its underlying tenor illuminated. Rather than a "passing into a foreign language with the help of one's own", translation is an "awakening, clarification, and unfolding of one's own language with the help of an encounter with the foreign" (ibid. 66). Alongside being a technical matter of substituting one language with another, translation is an encounter (*Auseinandersetzung*) with a foreign language, which sheds light on the meanings and nuances of the language of one's own. The other, the foreign, must avoid being assimilated, annihilated, or turned into an ideal, as such relations to the foreign would preclude its possibility to form a commencement to a deeper understanding of the own (ibid. 65–66).

Bearing this in mind, some elements of Heidegger's reading of Hölderlin's hymn and Sophocles' *Antigone* are of interest for this book, especially since *Antigone* is taken to express a relation and understanding of foreignness that diverts from the representational identity and calculability inherent to a German (or Western) thought confessing to the 'clarity of presentation'.

For the purpose of this text, Heidegger's close reading of the choral ode in *Antigone* is left aside. In focus instead is how Heidegger finds the essence of the human and of *polis* in groundlessness and questionability. In the *Antigone* encountered through the lens of Heidegger and Hölderlin, the human is an uncanny being. Her relation to the world escapes being pinned down by any representation. The human is manifold, not in terms of being various or many in kind, but as in being multiply *folded*—she is placed together, at once individuated and interwoven (ibid. 68). Extraordinary as she is, the uncanny is skilled in everything ordinary so that she also can rise above whatever is habitual. Heidegger soon highlights the relation between the uncanny (*Unheimliche*) of the discussion on *Antigone* and *polis* and the unhomely (*Unheimische*) of "The Ister" and finds reason to understand the former through the latter.

Keeping with the intrinsic equivocality of her essence, this homeless, uncanny being is the one who, without arrival, follows any path, while yet coming to nothing.[10] Why, if able to engage any journey, would the uncanny, homeless human being come to nothing? Heidegger writes:

> In human beings reaching everywhere, however, and in each case coming to 'something', they still come to nothing, because they remain stuck with particular beings in each case, and fail to grasp their being or essence in such beings. The 'nothing' to which they come is that which, turning counter to being, directly excludes human beings altogether from being. (ibid. 76)

Read in the light of the question of community, the reason the human comes to nothing in her relation to other beings is because she finds in the other a *particular* something—an equivalent, an identity with something of her own. By so doing, the possibility to initiate a return to her own is barred. One could say she is stuck where she started, with an image of herself projected onto the one she meets. The human cannot come to and appreciate the other, nor can she return to herself. In fact, she has never undertaken the first steps of such a journey to begin with. Therefore, Heidegger writes, "In those beings they come to, and in which they think themselves at home, they come to nothing. Thinking they are homely, human beings are those who are unhomely" (ibid.). In other words, when the human being is attentive only to points of identification, she is bound to err in homelessness. With identification, the human will either see in the other someone identified as same or as opposed or different. In the first case, the other is assimilated to oneself; in the second, the other is annihilated in the image of an opponent or other, which itself is constructed on the basis of oneself. Neither will suffice as a ground for a questioning turn to one's own. Despite her fervour to be "at home", the human is unhomely because she allows identification to stand in the way of her ability to grasp those she meets in their being.

With *Identity and Difference*, the last quote could also suggest that the human *does* make her way to an insight in her self and in being. That is, the human comes to someone in such a way that the one she comes to can prove a provenance for a return to the homely. Though what the human comes to in such an encounter is nothing. This nothing is not a mere void, but rather the impossibility to pin down any fixed content to the other. Here lies the seed for an encounter, taken in its full sense; that the search for *something* is given up and that instead the groundlessness of being is allowed to guide the meeting between beings, so that the human, guided by nothing, engages the other and herself through their common questionability.

Indeed, the human must learn to feel at home in *not being at home*. She must learn to feel at home in the groundlessness of her being; in a nothingness that still sparks a continued encounter with the other.

This is the task Antigone—the most uncanny of beings[11]—takes upon herself. She assumes as her own "that against which nothing can avail since it appears of its own accord, no one knows wherefrom" (ibid. 117), though this is nothing but being itself. Antigone reveals the counterturning character of the uncanny. It describes a homelessness where one is driven amid beings without reaching anywhere. At the same time, the homelessness of the uncanny consists in her "becoming homely from out of a belonging to being" (ibid. 118). If this remembrance of being is what Antigone takes as her ground, then she at the same time loses ground. The remembrance of being encompasses the insight that being itself is always questionable: "Being as being remains groundless. Ground—namely as ground which grounds being—stays away from being. Being is an abyss" (ibid. 111).

It is this uncanny being that the modern human needs to encounter if she is to release herself from the limitations her self-consciousness puts on her. Heidegger sees the self-consciousness expressed in modernity as certain of itself and of all beings it experiences. The self-consciousness of modernity makes everything indubitable and surveyable and sees how everything can be calculated and planned (ibid. 94). However, when the human relates to other beings as present-at-hand, she is in fact erring since this suggests that she can grasp something of a ground in herself as well as in other existing selves. Such a ground is illusory and will not disclose others in their being but as "manifested presence" (Odysseos 2003, 187). The latter cannot capture the indeterminate, contingent, heteronomous, and, above all, groundless being of other beings. A determinate and definite nature of the other or the self can therefore never be identified (ibid. 187–91).

This sustains the somewhat wilful reading of the lines from the choral ode of *Antigone*. Either the relation between the self and the other is based in the identification of a familiar—yet illusive—ground or their relation is formed through a difference, which still remains within the realm of sameness, since that difference is determined on the basis of the assumed ground of the self. In each case, the relation is a relation insensitive to the lingering disclosing, groundlessness, and questionability of beings. In this relation, the human comes to nothing, but not nothing in the sense of something slipping out of her grip when trying to grasp it, nor nothing in the sense of revealing the groundlessness and indeterminacy of Being. Instead, she comes to nothing by holding that something which she finds too tightly in her hands, by annihilating its being in that all too constrictive embrace.

The questionability of beings pertains also to the discussion of the *polis* in Heidegger's reading of *Antigone*. Further, the *polis* is given a spatial image and has, I suggest, some spatial implications.

Heidegger carefully points out how the *polis* cannot be determined politically. The opposite is possible; the "political" is something that belongs to *polis* and must hence be determined in relation to and in terms of the *polis*. To address the *polis* through the political cannot be done, he maintains. Just as we cannot decide the *logos* from what is perceived as logical or an *ethos* from a sense of the ethical, it is not possible to turn to the meaning and manifestation of the political in order to determine the essence of *polis*. If the political is what belongs to the *polis*—as the logical to the *logos* and the ethical to the *ethos*—and hence is dependent upon *polis*, then the essence of *polis* cannot be determined in terms of the political, since such a determination would be circular (Heidegger 1996b, 80).[12] But this does not suggest a prepolitical determination of *polis*. Heidegger's avoidance of a definition of the *polis* does not prevent him from asking for its essence. A definition, he says, might certainly point in the direction of something's meaning or essence. Yet what if that essence involves that which evades all definition and is worthy of question? Thus, Heidegger frames the essence of the *polis* as a suggestion: "Perhaps the πολις is that realm and locale around which everything question-worthy and uncanny turns in an exceptional sense" (ibid. 81).

Maybe, he says, *polis* names the site where everything becomes questionable anew. The *polis* is articulated in spatial terms; it is a locale and a site. It is also mentioned as a swirl (*Wirbel*) "in which and around which everything turns" (ibid.). As a swirl, the *polis* draws the uncanny human into its midst, so that she enters into its realm by keeping with the expanse of its limits (ibid. 82–83). As a locale, the *polis* is vested with flow, with constant change and movement. As seen in the previous section on dwelling, this flow with its inherent change and constancy does not simply emanate from the site itself but also from its interrelatedness with those inhabiting that site. This gives thought to an image of not only the *polis* as a site, but of the *polis* and its site as co-emerging.

Elden picks up on the thought of the *polis* as a site in *Antigone* in *The Birth of Territory* (Elden 2013a). For him, the *polis* is not only a people, but also the site where they live and where action takes place. However, this reading draws the site as something relatively stable and inert, as the backdrop against which the history of the *polis* is played out. As backdrop, the site can be modified due to the course of history, but it still assumes the role of a stage, as something acted upon, rather than itself being active in the staging of the *polis*. Indeed, Elden even finds reason to articulate the site as *anterior* to the *polis* when attending to an exchange between Haemon and Creon in *Antigone*:

> Haemon: The whole community [homoptolis leos] of Thebes denies it.
>
> Creon: And is the polis about to tell me how to rule?
>
> Haemon: Now you see? Who's talking like a child?
>
> Creon: Am I to rule this for others—or myself?
>
> Haemon: It's no polis at all, owned by one man alone.
>
> Creon: What? The polis is the king's—that's the law!
>
> Haemon: What a splendid king you'd make of a desert island—you and you alone.
>
> (quoted in Elden 2013a, 30)

To Elden, Creon's avoidance of the word *polis* in the last line of the above passage, as well as in other places in *Antigone*, is taken to show how Sophocles stresses the site as anterior to the *polis*. Instead of talking of the *polis*, Creon refers to the land (*khthonos*) or to the realm (*khoras*), something taken to emphasise the site, though not necessarily the *polis*. This is understood to be demonstrated in the last "retort" by Haemon from which Elden concludes "a desert island with a single person is no *polis*. It has the site, but no people" (ibid.).

Contrary to Elden, but along the lines of Heidegger, one could read Creon's avoidance of the term *polis* as if Sophocles is appreciative exactly of the *polis* as a site. What is described by Haemon, whose last line could read as also sardonic and uttered in scorn, is how the land corresponding to the '*polis*', when the latter becomes reduced to the rule of one man alone, is nothing but a desert island. When the manifoldness and questionability of the *polis* is lost in the word of the king alone, then also the swirling site of the *polis* is erased and whatever realm remains is to be equalled with a desolate island. This suggests not simply that the *polis* is "simultaneously a place and the people who inhabit it" (ibid. 31), but that the people and the place are *co-emerging*. The site is not the stage for the people or a land it inhabits. Rather, the site is *enacted* by the people; the place of the *polis* emerges with the ones who are present in the *polis*. And since those present in fact always are *becoming* present, since they are unfolding and emerging, it follows that the *polis* itself—as a site and as a people—is equally constantly unfolding and presencing.

PLACE FOR ALL?

So what would be required of dwelling beings? In the discussion on *Antigone*, being a dweller is tied to the questionability of existence. Even if this renders the appearance of anyone questionable, it does not necessarily withdraw from a representational thinking. A representation is questionable; it can be contested and augmented. It is also irreconcilably tied up with language, or, more exactly, discourse. This is no surprise, as it is in the density of speaking and thinking that Heidegger finds a remedy against the hold numbers and calculation have on existence (Elden 2006). In the case of the *polis* above, it is questionability that disrupts any settled meaning of the *polis* and paves the way for its turning augmentation.

This, however, makes the community emanating from Heidegger's thinking vulnerable to the same critique as directed to Rancière. How can the ones relying on a clandestine existence be seen to form a part of the community other than through being at the liberty of others to engage in speech? And in what way would the place of the community escape being caught up in representational, even if questionable, identity? Heidegger most certainly stresses the question, questionability, and language in a way that seems to lead right up to a community conjured through representational thinking. While parts of the next chapter address the gravity to which Heidegger's emphasis on questionability and speech is complicit in a community predicated on sameness, I here suggest that the incalculability of the spatiality stemming from the river undermines the very ground for Heidegger's question.

"Building Dwelling Thinking" presents the relation of place, landscape, and space as follows. With the bridge, there is also a location (*Ort*). The location, the place, is not there before the bridge nor does it come after, but the location is only a location because of the bridge. As a location, the bridge gathers a site that brings together the elements of the world. Such gathering (*Versammlung*) is a gathering in unity. Bringing together the unity that Heidegger calls the fourfold (the earth and heaven, the mortals and the gods), the site is already within an understanding of being. Allowing for a site, locations also open spaces. In "each case", the site "provides for a space" (Heidegger 2001a, 152). As a location, the building gathers a world as a site, a world from which spaces unfold so that the location comes to appear as an *Ereignis*. Heidegger writes how the location both *admits* and *installs* the fourfold. The place allows for a world to be brought together, and it brings this world together in a certain way so that both spaces and "space as such—'space'" are provided for (ibid. 154–55, at 154). Hence, space is already within the world, within "the stay of mortals" (ibid.). If to schematise, the location emerges as an event of appropriation, the site as where the plenitude of being is stabilised and rendered a particular meaning, and space as the spatiality emanating from whatever meaning of the site. This is why building is a founding and joining of spaces. As forming locations, building joins together spaces, though never as a single entity of pure space, but as the spaces emanating in the difference of how locations open for them to unfold (ibid. 156). Yet the location, the place that forms the shelter of the elements of the world (ibid.), requires the site for gathering these together. Maybe the need Heidegger shows to insert the site between place and space hints at a place that cannot be pinned down with a particular meaning. If that is the case, then place appears to engage both presencing and absencing in a thoroughly evasive manner, whereas the site emerges as where place gets caught up in meaning and representation. The latter aligns somewhat with Casey's interpretation of place as *preceding* the site, leaving place entailed in dwelling, whereas site forms the homogeneous space of a world of objects (Casey 1993, 177).

The river place is slightly different from dwelling place. If the poet, the conjurer of the poeticised river, for a moment is neglected, then the river lets a place emerge without human interference. It is not building or dwelling that is decisive for place but the river itself; "the river determines the dwelling place of human beings upon the earth" (Heidegger 1996b, 20). The abode, the dwelling place, has a locale of a special character, since "the river 'is' the locality that pervades the abode of human beings upon earth" (ibid. 21). It determines humans to where they belong and to what is their own. As discussed, for Heidegger, the appropriation of one's own is played out in the encounter between the Greek and the German so that historical unfolding, as well as the course of the river, is inscribed in one particular setting. Janicaud, however, encouraged a distancing from these determined nodes of history and suggested a reading set free from the assumption of there being merely one law of history. As fellow travellers to Hölderlin and Heidegger—but also as travellers drifting from these companions—we could instead attend to history as open, undecidable, and multifarious (Janicaud 1993, 391).

The river renders the place with both locality and journeying and attributes thus a movement and transience to place. With the river, the here is never really *there*. Place is painted as ephemeral and fleeting and as something that comes to pass but that also is to come. If freed from the precedence of the German-Greek, this journeying of locality rests in *any* encounter. Place is then not grounded in the historical coming of a re-encountered lost Greek origin. Instead, journeying gestures to the origins that lie in the encounter between *any* entities. That is, it gestures to how any entity has an embedded and intertwined relation to a place that is not constant but always under the law of the journey. Thus, place unfolds in the encounter between *any* beings, not only those that are part of a particular historical setting.

This would counter some of Heidegger's characterisations of the encounter between the foreign and the homely, especially the one seen in the lacking reciprocity in the relation between the home of the motherland and the foreign of the colony. To Heidegger, the journey into the colony is always a journey back to the "motherland" but can only be so if the motherland is first forgotten and hence can be discovered anew (Heidegger 1996b, 131). This assumes that the foreign *colony* can be seen through eyes cleared of any recollection of what they have left behind. The figure of the colony also lacks any engagement with its foreign and the consequent turn to itself, as if the fate of the colony, once turned into exactly a *colony*, can be forgetting of "the motherland".

When engaged predominantly temporally, journeying withdraws from the other, the colony. Indeed, the fate of the "colony" and whatever path its encounter with the "motherland" sets it onto become irrelevant. It is only the lessons drawn by unhomely motherland's encounter with the colony that matters in the course of becoming homely. "The boldness of forgetting in the love of colony is the readiness, while in the foreign, to learn from the foreign for the sake of what is one's own, so as to defer what is one's own to until it is time" (ibid. 132). Read spatially, however, the journeying seems not to come to an end. The journey not only transforms both sides; it draws them together, entwines the two and their fates. In the spatial image of journeying, Heidegger writes that the river

> flows hesitantly: such hesitancy can come only from there being a mysterious counterflow that pushes counter to its originary springing forth. … In such hesitancy the poet intimates the mysterious concealment of the intertwining relations toward the foreign and one's own. … In hesitating, its flowing appears in one direction *and* the other. The flowing is not directly in either of the two directions. The relation to the foreign is never a mere taking over of the Other. The relation to one's own is never a mere self-assured affirmation of the so-called 'natural' or 'organic'. (ibid. 143)

As flowing hesitantly, the river lets not one direction be given precedence over the other. Thinking through the spatial and with the river, there seems to be no distinct divide between the other and the self. Their separation consists rather in the long extension of the river that gathers them both in a stretched out here upheld by a hesitant flow that is also a counterflow.

Roberto Esposito warns against the dialectics at work in the appropriation of the own. He asks if the journeying in the foreign fixes "its point of arrival beyond the voyage", or if "the point of arrival coincide[s] with the journey itself" (Esposito 2010, 107). Esposito opts for the former and turns to Hölderlin's notion of the sea to grasp a community based in a continued sharing. I suggest that attending to the encounter between the own and the other through the river and through the spatial allows for the opposite conclusion. With an historical view, the encounter between the own and the foreign seems to depend on a boundary *between* the two. This boundary forms a threshold that allows for crossings over to each side to be intimated yet with both sides kept apart. Further, in the historical view, the journeying is fettered to the whole of an 'us' (Esposito 2010, 107; see also Schatzki 2007, 58–59). Spatially, in contrast, the journeying of 'us' can equally be read on the level of each being. The spatial view of the own and the foreign draws thereby less a boundary than it inscribes a bond between beings, a bond that is rendered an image in the hesitantly flowing river. Coming to oneself, appropriating what is one's own, takes here the form of an oscillation back and forth between the self and the other, so that they encounter each other in their presencing and withdrawal. As such, they must precisely *not* be appropriated by each other, neither must their own be fully appropriated by themselves. Hence, coming to oneself, with the hesitating flow of the river, suggests a journey that cannot be completed and hints at a community in, as Esposito puts it, "our shared inability to be appropriated" (Esposito 2010, 109).

Still, there is an ambiguity to the spatiality unfolding from the river. The river Hesse's Siddhartha sits down by and through which he sees himself and all forms as transitory seems to share a secret with the Ister. The river is *there* yet always shifting. Siddhartha "saw that the water continually flowed and flowed and yet it was always there; it was always same and yet every moment it was new. (Hesse 1951, 100, 102, at 102). While Hesse describes the river as there, as always both same and new, I would rather say that the river is there—it is present before us—but as a presence that always escapes us, that is never merely there. As a place, the river is there yet always different. The river summons a peculiar place that is both present and different to itself at the same time.

In "Art and Space", Heidegger writes of how the realisation that there is "nothing" behind the phenomenon of space to which it could be traced back induces anxiety (*Angst*) in humans (Heidegger 1973, 4). It is not a threatening 'there' that evokes anxiety; rather, "the fact that what is threatening is *nowhere* characterizes what anxiety is about" (Odysseos 2003, 189). Anxiety interrupts the self-conception of the autonomous subject by reminding the self of its lack of any definite or determinate substance, "by reminding it that it is an entity whose Being is an issue for it" (Heidegger 2010, 180). Nowhere means not nothing, Heidegger asserts; for the "essentially spatial being-in" nowhere lies in the general region and in the disclosedness of the general world (ibid.). With anxiety, we are opened up to what we are, how we are thrown, and our fundamental possibilities. "Art and Space" articulates how men are overcome "by an awe to the point of anxiety" by space, not only since behind it, "nothing is more given to which it could be traced back", but also because "before space there is no retreat to something else" (Heidegger 1973, 7).

Just as such space, the river is on the one hand present, there, and always before us. On the other, it is a presencing and vanishing that evades any determinations in terms of objective presence. The only thing given in river place is that it is there, but this 'there' is itself never the same. River place, therefore, appears to describe a *difference that does not refer to an underlying notion of identity*. Any self-identity to the 'there' of the river means that its flow would come to a halt. If the river prevails over the essence of the locale, this also affects the latter. The essence of the locale, its locality, is an "interplay of places" (ibid. 10–11). Also place appears to show an irreducible alterity in an interplay both in relation to other places and to itself. Such interplay does not describe a continued resignification, but how that which is there, before being caught up in meaning, is itself both open and an invitation to openness. Place has *dunamis*, as Elden writes with reference to Heidegger; we can only think of place because we can think of motion. The *dunamis* of place gives way for a "proper presence" in opposition to geometrical objects that are externally determined in advance and have no place (Elden 2006, 126–27).

The above grasps place as inherently relational, leaving the question of the place of those out of sight and hearing open. Hence, is the world-building place available to anyone? Is the clandestine or solitary human, or, for that matter, the animal, able to dwell and be in place? While not writing much on the body, Heidegger mentions how the place (*Raum, topos*) is formed through the body and shares its limits with the body. Yet again the limit is characterised as a site for emergence, now with the added meaning that the place begins with the body, so that every body has its own place. The human relation to place, in turn, is such that it opens and releases place through letting herself into it. And when in place, the human being is also part of a world (Heidegger 1996a, 10–15). From this, it appears that place does not necessarily mark a relation unfolding from two beings, but rather the relation between the human, bodies, their place, and their world.

This resonates with another poetic encounter with space, in which Gaston Bachelard writes of the world-building qualities of intimate and solitary places. Bachelard considers imagination as "a major power of human nature", since, through imagination and daydreaming, worlds not curbed by the obstacles of reality can unfold. But daydreaming is closely related to intimacy; we daydream in the places of solitude that we can carve out for ourselves (Bachelard 1994, xxxiv). The places where Bachelard sees worlds emerge are not public squares or grand construction projects. Rather, he turns to the places where one huddles up—to the intimacy of the nooks and crannies that provides places for solitary daydreaming. When asking of the poetics of space through the intimacy of the home, Bachelard does not refrain from reaching beyond the human realm. Nests and shells are places of refuge, where the creature "takes to cover, hides away, lies snug, concealed" (ibid. 91). This makes them no less part in bringing forth a world. The bird adds to the nest with sticks and feathers and forms it by the pressure of her body; the shell is alive: it is a "house that grows with its inmate" (ibid. 118). Bachelard's contemplations on chests, caskets, and drawers all serve to open up a new dimension—the dimension of intimacy. It is only from this dimension that the cosmos of daydreams unfold. To Bachelard, then, the grandiosity of buildings such as the bridge is exchanged for the intimate corners of the world, from which spaces open up and unfold toward new thought worlds. Intimacy, to Casey, withdraws both from calculability and from language and representation. It forms a link between place and self that cannot be established through any objective measurement since these would be superimposed on the experience of place. Intimacy is not necessarily mental but foremost a bodily experience. Place is not logical or metaphysical but phenomenological and felt—"felt bodily first of all" (Casey 1993, 311–13, at 313).[13] Hence, through the notion of intimacy, it is not whether a being can step out in the public or address the world as such that determines if she can gather a place around her. Quite the contrary, she can dwell in the intimate, speechless and out of sight, and still be part of the space unfolding from the place she gathers by doing so. Next to the freedom inherent to the place of dwelling there emerges also an equality suggesting that neither material circumstances nor intellectual abilities are required for world building.

Thus, place is not bounded; place is incalculable, and place is for all. There is place for all. Being a difference that does not rely on identity, place suggests that not one body is in place in the same way, not with respect to other bodies, not with respect to itself. Still each one is world building. Prevailing over the essence of place, the river has already made of place an interplay of places and put in play the difference that place is.

ALREADY PART

Place, as elaborated above, appears to challenge representational thinking. Jean-Luc Nancy writes of how Heidegger abandons the concept of freedom for the openness of the "free" or the "free space" (Nancy 1993, 40–41). This free space is peculiarly exempted from identification, which means it is also withheld from questioning. It seems to disrupt both truth and what it conceals and sets these free in a play deserting the mastery of language.

The notion of place emerging here has implications for the non-included. She can no longer be seen as a discrete phenomenon within a certain territory. She forms not an isolated event ready to be undone through her removal from the territory of her presence. Instead, she forms an intimate and already integrated part of the place in which she emerges. Her presence alters the site where she dwells, precisely due to her dwelling. As Gregor Noll concludes, "The emphasis on dwelling unconditionally presences those undocumented migrants who have found a home, however precarious, on the territory of a host state" (Noll 2010, 253). This deviates from Kant's "right to visit", which proceeds from the shape of the Earth as a globe and the fact that humans dispersed around this globe eventually will meet each other. From this derives a freedom of movement that allows for the migrant to appear in any foreign territory, whose population has the right to ascertain her removal. Noll therefore concludes that Kant "keeps the migrant eternally moving. She can only appear in the capacity of a *migrating* human, a human characterized by her propensity to disappear—or to be made to disappear" (ibid. 249–50, at 250). Thus, also appearing as *something*, she can be made to disappear. What appears is *presented*, as if bringing into daylight something already made. Thereby she fits into a scheme; the stateless is made to map onto an already existing category that excludes the singularity of her being. Hence, she can also be made to disappear while the category—the same to which she is understood to belong—remains as a category that harbours the hospitality of those included, while simultaneously conditioning this hospitality on the difference of its objects.

In contrast, as dwelling, the formally excluded forms an undeniable *part* of the abode of the community. Her dwelling is already immersed in the landscape and takes parts in its animation. Through this perspective, she cannot remain excluded due to her potentially being in a position where she is not at liberty to articulate an account of *logos* that becomes appreciated as such by the consensus community. Instead, her place gathers space as incalculable, as a free space without mastery. As Nancy points out, a "total liberation of freedom" could not "occur as the total mastery of freedom" (Nancy 1993, 147).

Instead of seeking to understand how the way we understand action, political existence, and belonging is determining the way we grasp the space of the political community, this chapter sought to turn the question of political community around. How, it asked, does the way we think about the spatiality of the community determine the ways we think about coexistence, political action, and belonging? Returning to the problem of the stateless, it can now be concluded that place as a difference that rejects any identity is of manifold significance for the ones formally not part of a community predicated on sameness. *First*, place does not demand of the dweller an ascription to a particular account of sameness set up against difference. On the contrary, the character of dwelling is found in sparing, which means to leave something free and at peace. This, it is suggested, must include being freed from the mastery and subjugation of representational thinking. *Second*, and in consequence, dwelling unconditionally presences the formally not included as an intertwined part of the place where she resides, since she also takes part in its gathering. By being there, by dwelling, the non-included is already part of the place through participating in it being conjured. Read together with the Ister, this means that by her presence, the non-included has already altered the character of the abode of the community, as well as its historical unfolding, present as well as past. *Third*, since place is always different and lets the place of the community unfold along with the ones coming to presence in and with that place, it follows that the human abode can never be pinned down as self-identical. The community takes place in an incalculable spaciousness in which place is intimately entwined with movement and change. Such a place impedes any notions of an emplacement or re-emplacement of displaced subjects. Those measures would sustain difference between included and excluded as a division to be transgressed, yet leaving the distinction between inside and outside in operation. Instead, presencing along with locality and journeying suggest the place of the community to exist precisely in its incalculable disclosure. As presencing, the abode is rendered a spatiality sensitive to how each place brought forth by every instance of dwelling is already an intertwined and constitutive part of the community. Thus, the spaciousness of presencing is perceptive to how *anyone* is actively participating in the gathering of the place of the human abode when being in, and of, place.

If belonging, then, stems from one's existence as being in place, it cannot by identity claims emerge as a question of membership, as such forms of boundaries would contradict the very foundations for dwelling and place. Likewise, coexistence cannot be predicated on a notion of sameness presupposing difference. On the contrary, coexistence stems from our shared inability to be appropriated. Through the spatiality of a place without identity, one can start to discern the statefree, for political action is no longer directed towards a particular *logos*, or a particular understanding of what we perceive as the world. To act is no longer related to a particular script. Rather, to act is to be in place, and as such, it is already a joint venture into the unpredictability of being together.

But what does this do to the question of the *political* community? If statelessness was symptomatic to a particular meaning of politics structured around boundaries, how could the political be understood in relation to the incalculability of place outlined above? And how would the political be understood with respect to the statefree? The above elaborated notion of place undermines not only the individual and identity. If the individual is inherently connected to the sovereign state, then also the latter is challenged by place. So is also a notion of difference mediated by representation. The political implications of an incalculable space and of place as difference without identity are immense. The incalculability of place and a free space stands in stark opposition to the territory and the space of sameness, which characterise a space of authority. Incalculable place cannot be mastered without it turning into the sameness of ontic presence. Hence, sovereignty and representational thinking, as well as a community predicated on sameness, seem to stand and fall with the place of place in the thinking of political community. However, before venturing into these questions, a deeper insight in the relation between incalculability and rationality needs to be gained.

NOTES

1. "Heidegger's term is 'Wesen'. It is used in the verbal meaning of φμσις rather than the more static meaning of nature or essence" (Stambaugh 1969, 14).

2. These are the two key expressions in the lecture on identity and difference, and they are also so in general in Heidegger's thought. Therefore, the translator's note on the terms are quoted at length: "'*Ge-stell*' in the sense in which Heidegger uses it does not belong to common language. In German, '*Berg*' means a mountain, '*Gebirge*' means a chain or group of mountains. In the same way '*Ge-stell*' is the unity (but *not* a unity in the sense of a general whole subsuming all particulars under it) of all the activities in which the verb '*stellen*' (place, put, set) figures: *vor-stellen* (represent, think), *stellen* (challenge), *ent-stellen* (disfigure), *nach-stellen* (to be after someone, pursue him stealthily), *sicher-stellen* (to make certain of something). The event of appropriation (*Ereignis*) is a word belonging to common language and means 'event'. But Heidegger's use of it is more (1) 'abstract' in the sense of being infinitely removed from everyday events and yet of being that which is so close to us that we cannot see it, and (2) 'concrete' in its use of the very *roots* of that word: *er-eignen* (*eigen* = own, thus to come into one's own, to come to where one *belongs*) and *er-äugnen* (*Auge* = eye). This is the real etymological root of *er-eignen*), thus to catch sight of, to see with the mind's eye, to see face-to-face."

3. Attentiveness (*Aufmerksamkeit*) is an important term for Heidegger. Attentiveness is awakened by a "thoughtful reflection" and takes its form in a "fundamental attunement" to the essential. Attentiveness marks out "the essential from everything else so as to retain it in the future" (Heidegger 1996b, 13). This is different from an interpretation of a poem, which would draw on the knowledge of the reader. Attentiveness rather operates as a de-parting, as an initial attempt to clear a space for thinking.

4. Interestingly, it is partly through meditations on the river that geometry begins its consolidation as a measure of space—or at least of land—when Euclidean

geometry is merged with Roman law in the fourteenth century. In his treatise on Tiberiadis, Bartolus de Sassoferrato discusses issues pertaining to the river Tiber. He especially considers how to divide alluvial depositions and islands that emerge out of the river and also how to divide an abandoned riverbed. Here, geometry offered a possibility to move beyond peasant knowledge and distribute alluvial land in a more fair way (Cavallar 2004). Also Hugo Grotius and Emer de Vattel discuss the river in relation to legal issues (see further Grotius 2000, 10–8; Vattel 2008, 242–49). Here, the emphasis is less on geometry but more on the river constituting a natural border. This contrasts to Heidegger's account, in which the river rather appears as something of a *link* between a there and a here. For Grotius and Vattel, the opposite is rather the case: rivers tend to pose natural boundaries between territories; "nothing can be a more apt mark of distinction than those impassable bounds assigned by nature" (Grotius 2000, 107) and "nothing is more natural than to take a river for a boundary, when a settlement is made" (Vattel 2008, 244). Still, whatever alluvial deposits that come out of the river belong to the owner of the riverbed, but the river itself—in it being running water—is common to everyone (ibid. 323–24). Lauren Benton, finally, also sees how geographic and political-legal imaginaries become linked through rivers (see further Benton 2010, 40–103). She studies how rivers have served the imperial expansion for colonial powers when forming routes of quest and conquest into the unknown. Rivers were central for controlling distant subjects as "sovereign spaces followed [the] corridors of control" made up of river networks (ibid. 104).

5. Malpas maintains how "'Event' in itself is topological in character—perhaps *the* topological concept in Heidegger's later thinking" (Malpas 2006, 214; see Raffoul 2012a for a discussion on the event of space).

6. And the time belonging to that life, I wish to add, bearing in mind de Beistegui's critique of what he sees as a downplaying of time and history in *Heidegger's Topology*. De Beistegui's discussion does not unfold around the theme of death as limit and the finitude of place. It raises instead the concern of Malpas not giving sufficient thought to how his work is situated also historically and how the question of place, as Malpas understands it, only makes sense in a historically defined moment (de Beistegui 2011).

7. Young here foregoes Albert Hofstadter's translation of *schonen* as *sparing* and *preserving*. Instead, Young suggests 'caring-for' as a more suitable translation with the stated reasons being that caring-for (a) corresponds to Heidegger's use of only one word, *schonen*, and (b) better "captures the creative, 'freeing' aspect (in other words *techne*)" which Young argues is an essential aspect of *schonen* (Young 2002, 64). Presumably, caring-for is also preferred by Young due to Heidegger's central notion of 'care' (*Sorge*) in, especially, *Being and Time*. Nevertheless, to translate *schonen* as caring-for means to give a more emotional charging to the term. This move makes *schonen* more suitable for a discussion of ethics, but it also turns *schonen* into a practice invested with a concern for and attachment to its object. *Schonen* then stems from an active engagement with, for example, another being. This, nonetheless, neglects the more disengaged meaning of *schonen*—the meaning of letting be or leaving at peace—which does not necessitate the same degree of attachment, if any at all, as does 'caring-for'. While not the object of discussion here, it appears as if it is

precisely by omitting the more detached elements of *schonen* that Young can appreciate dwelling as a predominantly ethical concept and dwellers as the guardians of the world we all inhabit.

8. *The Ister* is in fact to a large extent omitted from Young's discussion, a move most peculiar since dwelling and man's homelessness (or unhomeliness) is a dominating theme in the text.

9. The uneasy feeling that this image invokes is by no means relaxed when Heidegger later lifts the colony as a place for forgetting of the motherland so that one can return to the latter with eyes opened from the journey into the colony (Heidegger 1996b, 131).

10. The quote from the choral ode in *Antigone* reads: "Everywhere venturing forth underway, experienceless without any way out he comes to nothing [*Überall hinausfahrend unterwegs erfahrungslos ohne Ausweg kommt er zum Nichts*]" (Heidegger 1996b, 73).

11. Heidegger does not hesitate to place Antigone in the heroic character from which Honig sought to retrieve her.

12. Leo Strauss makes a similar observation when noticing the impossibility of defining the political without a prior understanding of what constitutes a society (Strauss 1959, 22).

13. Taking issue with the centrality of language and the word in Heidegger, and how his philosophy thereby remains within the meta-physical *logos*, Luce Irigaray invokes the spatiality of *air* as the "'intemporal' spatiality where everything takes place" (Irigaray 1999, 167). Through emphasising air, Irigaray sees the possibility to remind Heidegger of "his own insight that language is rooted in the body" (Fielding 2008, 221). Air is the materiality most near to beings, but contrary to language, air does not impose itself as a verbal construction on the world. Instead it forms a possibility to approach the human beyond the insufficiencies of language, and especially through its withdrawal from any delineating appropriation. Without appearing, air is present and constitutes the means for communication and hence also for forming relations (ibid. 221–24). It is especially a particular body and relation Irigaray engages to open a space for thinking: the body of the pregnant woman and the relation between the mother and her unborn child. These figures allow for an altered spatiality. With respect to Heidegger, the dark place of the womb lends thought to a wordless dwelling in nearness, yet without assimilation. In a mimetic reading of Maurice Merleau-Ponty, the darkness of the utero gives precedence to the tactile over the visual and the invisible over the visible. In relation to both thinkers, the priority of sight is downplayed. The tactile positivity in the darkness of the uterus reverses the priority of sight over the tactile that Irigaray attributes to Merleau-Ponty (Irigaray 1993, 152–54, 164–65). Also in Heidegger, she is at odds with a letting-appear that imposes a way of seeing which is complicit with already known categorisations and recognitions (Fielding 2008, 224). Instead, the word- and sightless dark place of the womb would serve to redirect existential thinking to the prelinguistic, to the felt, and to the bodily experienced.

Chapter 5

Reasoning with the Incalculable

The previous chapter explored the meaning of incalculable space and place as a difference that escapes identity. It was seen how incalculable space appears to undo the hold calculability, mapping, and sameness have on space and interrupt the image of bounded territories and space as continuous extension. Incalculable space and place, it was argued, will presence anyone emerging with a place as an integrated part of the place of the community. Place unfolds prior to representation; by gathering place, a being is already part of the staging of the site of the community.

Still, other attempts to engage with incalculability have been made by Heidegger, as well as by readers of Heidegger such as Nancy and Derrida. For the three of them, access to the incalculable is gained through the assumed possibility to *reason* with the incalculable. This chapter will explore some of these attempts. It turns to Heidegger's own remedies to the problem of calculability and discusses how his emphasis on language, questioning, and humanity eventually delimits his existential philosophy to encompass a chosen few. This discussion is followed by Nancy's and Derrida's attempts to get out of Heidegger's impasse. While they provide indispensable attempts to open up the difference of language, it is argued that their reliance on reason has exclusive consequences. Engaging the same thinkers, the last section expands on the relation between the calculable and the incalculable in order to give way for some of the concluding discussions.

First, however, the chapter seeks to draw some conclusions from the previous discussion on dwelling and the problem of how we must ever learn to dwell. This is read as a predicament of being caught between the incalculability of the lived experience of place and a *logos* ordering the world and its appearances in accordance with representative and calculable thinking. Here, it is suggested, emerges a venue for grasping a notion of sameness that does not presuppose difference.

SAMENESS AS A PREDICAMENT OF SITUATION

The previous chapter touched upon how the *polis* in *Antigone* was grasped as a site. It argued that this site emerged and unfolded with its people. Like the river draws an indeterminate human abode, the co-appearance of the *polis* and its people forms a site of continued emergence. In a discussion of Heidegger's conception of *polis* as a site, Elden suggests that Heidegger's account of the *polis* opens an understanding of politics not as conflict but as situation. Elden sees Heidegger distancing himself from Carl Schmitt and the risk of "confusing the political with the polemical" (Elden 2000, 416). Heidegger is taken to offer a parting from Schmitt's characterisation of the concept of the state as presupposing the concept of the political and its distinction between the friend and the enemy. Instead, Heidegger describes how the concept of the political presupposes the concept of the *polis* (ibid.) and correspondingly, as I understand Elden, how the situation presupposes the site. What appears to follow upon the emphasis of the *polis* as site, then, is a notion of the political as *situation*.

Now, *situation* is one of the many concepts Heidegger engages. The situation (*Situation*) describes an existential orientation. It is not something already present but a *there* disclosed through the mode by which *Dasein* orients itself to being and the world. In *Being and Time*, this disclosing mode is found in resoluteness (*Entschlossenheit*) (Heidegger 2010, 286–87). While the term *Entschlossenheit* was understood as resoluteness in *Being and Time*, it later comes to indicate *Dasein*'s self-disclosure (Ruin 2013, 364–65). Hence, it is in and through *Dasein*'s own disclosedness that the there disclosed in the situation can be encountered. However, situation is not the ordinary comportment to the world, at least not in modernity. This is instead characterised by the *position* (*Lage*) which describes "an objectively present framework in which Dasein occurs or into which it could even bring itself" (Heidegger 2010, 287). The *Lage* would be the situation of the 'they' whose being is not directed towards being, disclosure, and presencing. In fact, "*for the they, however, situation* [*Situation*] *is essentially closed off*", all the they knows of is the "*general situation* [*'allgemeine Lage'*]" (ibid.). Situation as an existential orientation, then, is not something immediately available to *Dasein*. On the contrary, the general character of *Dasein*'s spatial existence is to simply be posited in what is perceived as objective space. Though as being posited, *Dasein* is not really *there* at all; there is simply no *Da-* of *Sein* to be had. Instead, the situation is predicated on the existential orientation of resoluteness or disclosedness, though not in the sense of a representation or an acknowledgement following upon resoluteness as being-in-the-world. Rather, to be resolute means to already be placed in the situation (ibid. 286–88).

When Elden lifts the situation as a possibility to grasp the political, he proposes that Lefebvre's suggestion how "there is a politics of space because space is political" with Heidegger could be grasped as "there is a politics of space because *politics is spatial*" (Elden 2000, 419). If the world is encountered through calculability and space through measurement, there appears to be no room for situation. Any *there* of a being would instead be marked as a mere position. Further, "resolute, Dasein is already *acting*" (Heidegger 2010, 287). Thus, resoluteness and situation are concurrent, as are, it seems, place and act. Alongside the co-appearance of the *polis* and the site emerges a *Dasein* disclosed in the situation, as well as in the act and the place.

The political as situation opens ways to think both Antigone's and Ismene's deeds, slightly different from Honig's. In Heidegger's Ister lecture, the act of Antigone's burial of her brother takes on a most poignant character. Her act takes the incalculability of the eventfulness of being, that is, "that against which nothing can avail as the point of departure governing everything" (Heidegger 1996b, 102). Heidegger writes that Antigone is in conflict over "the site of all beings", indeed, that she "steps out of the site altogether" and is so "utterly unhomely" (ibid. 103). Reading her deed spatially, Antigone appears to step out of the site of the *polis*, of the site as given by the edict of her brother, and of the site sung by the Theban elders. Though if the burial is a situation, then her stepping out of the site is simultaneously a gathering of a site; it brings forth the site. With her act, the site unfolds and emerges *anew*. Further, as gathering of site—as a situated staging of *polis* through a banned funeral—it seems like *both* Antigone's loud burial of her brother in broad daylight *and* Ismene's stealthy act carried out in the dark of the night must be registered by the *polis*. As situations, both acts force their way into the *polis*; both burials take place and are situated circumstances within the *polis*. Both the sisters emerge as the uncanny beings belonging to the *polis* by their situated deeds that again render the *polis* worthy of question. Stepping out of the site of the *polis*, they turn to the *polis* with a question that takes the form of a situated act, of situation. As a question, situation turns right towards the *polis* as a vortex and will cause it to turn and to expand its limits. As situation, the question is something over which no human being can "neither rule nor dispose" (ibid. 101); since it is there, it is already present in its presencing of a site gathered anew.

The previous chapter suggested dwelling to unconditionally presence the formally non-included as an intertwined part of the place where she resides, since the non-included also takes part in the gathering of its space. By being there, by dwelling, she is already a part of the place through participating in it being conjured. However, while we are all thrown into the world, the distinction between situation and position suggests that whither we are thrown can differ. Being posited indicates that we are thrown into a world ordered through calculability and categorisations and that our thrownness would be in full accordance with a rationality of calculability. Situation designates instead the activity of experiencing ourselves through a place conjuring a *there* in the world. This latter thrownness feeds not off the order of calculability, but rather on the indeterminacy of a place that evades all demarcation and ordering.

For Heidegger, the problem with dwelling is its scarcity. In modernity, the event of appropriation is forgotten—the eventfulness of the event is neutralised—and our capacity to dwell is sealed off in an orderly thinking that in advance has locked space and time into a grid of calculability that leaves room for nothing but what it manages to grasp. Modernity is the time of human homelessness and destitution. In this, the nature of dwelling can be searched for, but not once and for all found. This is where "the real plight of dwelling" lies—in that humans "*must ever learn to dwell*" (Heidegger 2001a, 159).

In its most proper sense, which for Heidegger would be in pre-Socratic Greek thought, dwelling suggests a relation to the world that is gathering place and allows for the presencing of beings in their being. Such dwelling means that one realises the groundlessness of being and finds oneself at home in the homelessness that such a realisation inevitably must lead to. However, the problem is that *Dasein* "is and has to be" (Heidegger 2010, 272). *Dasein* cannot escape making sense of things; the event of appropriation finds its appropriating meeting point in *Dasein* and in the mode by which *Dasein* is there in the world. This is why we must ever learn to dwell; to engage in a presencing of place and being, we must continuously question what unfolds from our sense-making. In the position, the *Allgemeine Lage*, the insight of us as sense-makers is forgotten, which means that our questions are only raised within the frame of calculability. Situation implies instead a *Dasein* attuned to her world in a questioning mode towards what is disclosed as well as what remains undisclosed and secret in this world. This means that *Dasein* is attuned to *how* her situation is gathering a place and how that place is not exclusive of other possibilities of being in place.

Thus, "situation" is not a noun, describing a set of circumstances, or a state of affairs. Instead it designates a coming to be in place. While place is a mode of being in the world *prior* to ontic determination, it is also that which is lost to a representational thinking or calculable knowledge. Situation aligns with Casey's notion of "*getting back into place*", that is, getting back to the experience of place (Casey 1993, 309). Situation describes the happening of, or the learning to dwell in, a place that always escapes any determined meaning or measure.

With the situation, it is time to return to the question of sameness. For in the tension between the position and the situation, I suggest there is a sameness to be found—*a sameness that does not presuppose difference* and that *we all* are exposed to and have to engage, at least as far as we are affected by the reach of modernity and its production of space. *This sameness consists in a predicament of situation.* Given that we are affected by modernity and its calculability—be it through living in it, with it, or against it, or through being subsumed under it through the reach of its power and control—we are also torn between the calculable space of the *Allgemeine Lage*, with its positioning and categorisations, and the unfolding place through which the situation opens for a dwelling attuned to what is disclosed, as well as to the lingering concealment and incalculability inherent to that disclosure. Dwelling is not something that we simply have or know; coming to one's own is that which is most difficult and an achievement ever needed to be undertaken anew. In this respect, there seems to be no difference between the ones resting assured in the assumed insides of the community and the ones at its margins or in the realms outside of formal inclusion. They all share in the predicament of being torn between being posited in a calculable position and being in a situation of disclosure and presencing. Thus, the predicament of situation consists in the tension between a modern positioning and a situation attuned to the presencing of being, in which a releasing from the former and a reaching of the latter is anything but easy. Engaging withdrawal and the forgetting thereof in this sense means that the predicament of situation draws attention to the presencing and unfolding of space and time in a frame of calculability where situation always to some extent is found wanting.

Sameness as a predicament of situation is here suggested as a sameness that does not presuppose difference or brings together the same under the sign of semblance of a particular representational identity. If situation opens to the disclosedness, presencing, and indeterminacy of existence, then the predicament of situation points at the difficulty by which this can be attained in a world of calculable reason and space. While 'we', whoever this is taken to entail, might not be exposed to this sameness in the same way, the predicament of situation emerges as shared as far as one stands under the spell of modern calculability and ordering, and as far as one appears within modern space and its *logos*. While this 'we' does not necessarily amount to a community, the predicament of situation has implications both individually and collectively. It has implications for the individual since the individual is disclosed only in the dwelling that situation opens towards. But since the 'I' has meaning only to the extent that it is related to an other, the predicament of situation has implications also collectively. One encounters oneself in the encounter with the other, but only as far as the two are kept apart; any assimilation or appropriation under a position will exclude the possibility of a proper encounter.

While the common experience of sharing in a predicament of situation describes a sameness anyone exposed to calculability, modernity, and representational thinking would be part to, it does not necessarily answer the question of the *political* community. Rather, it describes the existential predicament of sharing in a scarcity of being, presencing, and disclosure. Still, in *Antigone*, the *polis* and its site were seen to emerge together with its people. Taking the predicament of situation seriously as an existential predicament for all to battle has implications for the understanding and constitution of the community. *First*, to engage the predicament of situation as a common trait for modern beings suggests that one needs to take seriously the orientation to the other as a "letting dwell". By leaving the other at peace, by setting her free into her own presencing without her being assimilated or annihilated in a particular frame, the other might actually form a point of departure for that encounter necessary for one's own unfolding presence. *Second*, the predicament of situation means that we are all exposed to the risk of being inscribed in a calculative thinking that excludes us from being attuned to ourselves, others, and being as something unfolding and full of incalculable possibilities. To the extent that one at all finds calculability problematic, to the extent that one sees it as diminishing existence, then it seems as if a freedom-oriented politics implies an orientation towards the other invested by letting dwell, of the opening to and of place. And *third*, this would exclude any walls built around the community, since these would serve to press the communal site into territorial containers ridden with the forgetting of the manifoldness of existence. It would form a closure and a grounding that would preclude any presencing altogether.

HUMAN BEING

Approaching the question of the political community through its spatiality means to engage the situational character of the political community; that is, to grasp it as a site continuously gathered anew through its presencing. However, whereas my reading of Heidegger has predominantly circled around questions of space with the attempt to approach the notion of a community not mediated by conceptual identity and ontic representation, this ambition rhymes in part poorly with Heidegger's own reliance on language. I proposed that the lived experience of place is a world-forming activity stemming just as much from the intimacy of the bird's nest or the fox's lair as it emanates from public buildings and constructions. Indeed, I suggested intimacy and the lived, embodied relation to place to partake in letting a world unfold from a wordless place of incommensurable difference. For Heidegger, however, it is speech and poetry that can help release thinking from modern calculability and measure. It is also speaking, as well as listening, that guides being with one another, making Heidegger's community one of speakers and listeners (Elden 2006, 40). This section studies some of the consequences of the emphasis on speech by turning to Derrida's reading of three interlaced Heideggerian themes: humanism, language, and Western historiality. It is argued that these entangled threads sustain a sameness that narrows down the scope for inclusion, by installing difference as a necessary support for Heidegger's existential analytics.

Humanism, language, and Western historiality all tie into the question of being. Despite targeting humanism for ascribing a universal essence to the human and forming a metaphysical philosophy, the human is the protagonist of Heidegger's thinking of being. What is proper to man is the thinking of the question of the truth of being. Thus, man and being are inseparable, Derrida argues, and he unveils how, through the notion of a certain proximity—to oneself and to being—man takes centre stage of the drama of existence. Being is what is nearest to man of all beings, meaning that "Being is *what is near* to man, and that *man is what is near* to Being" (Derrida 1982, 133). To Heidegger, there is a co-propriety of being and man to the extent that the two are inseparable. This should not be understood as an ontic inseparability, which would amount exactly to the identity between being and beings that had to be deconstructed in Descartes's substance. Instead, the inseparability relates to *meaning*; the meaning of Being is related to the meaning of man. Indeed, what is proper to man, his *Eigenheit* or authenticity, "is to be related to the meaning of Being" (ibid.).

So while Heidegger seeks to contest a *humanitas* that is either metaphysically grounded or made the ground of metaphysics, he himself installs the human as the one who engages in the thinking of being. Derrida ascribes two contradictory moves to Heidegger. On the one hand, Heidegger sees all metaphysics as humanistic since metaphysics has already interpreted being on the basis of man in their forgetting—knowingly or not—of the ontological difference. This means that the essence of man is threatened by any metaphysics, and consequently, it seems, by any account of *humanitas*. On the other, since man is the one being who engages in the question of the truth of being, indeed, who finds his very essence in this engagement, Derrida sees a "revalorization of the essence of and dignity of man" (ibid. 128).

Hence, the thinking of being is a thinking carried out by man and man alone. In addition, the thinking of being is also a thinking of man, since the meaning of being will direct thought to the meaning of man. To ask of the meaning of being, or the truth of being, means namely to engage in *questioning*, which is a privilege reserved for man. We have already seen Antigone, the most uncanny of beings, engage in this questioning of what is foreign to her. She would not assume the foreign as merely opposed, but her engagement with the other rests instead in the insight of the groundlessness of herself and of the other in their being. Any relation that forms an encounter with the other is guided by the question called forth by groundlessness. This is what makes her belong to the *polis*, a *polis* into which she is admitted through a questioning that is "keeping with the expanse" of the limits of *polis*, namely, its questionability (Heidegger 1996b, 83). Already here emerges a bond between being, questioning, man, and the community; a bond, which also hints at a boundary, since, as discussed with respect to agonist political theory, questioning or contestation require a boundary that can be questioned, contested, and potentially overcome. This was the problem identified in Rancière's distinction between *phone* and *logos*. This distinction formed a boundary between those with power to determine if mere noise was to be perceived as voice and those whose contestation and thus also equality and liberty was conditioned on being heard by those with the power to hear or not hear.

This problem is also Heidegger's. At least it is a problem of his humanism, valorisation of language, and Germanocentric account of history, which all draw upon a privilege given to the question. Questioning, *Fragen*, the possibility to question, resurfaces again and again in Heidegger's *oeuvre* as that to which he ascribes what is highest and best in thinking. Questioning is the very task of thinking, and thought engages the question of being through questioning. The only thing that remains unquestioned, Derrida concludes, is the privilege afforded to the question. Although all must be open to questioning, the very possibility of questioning itself is withheld from being questioned (Derrida 1989, 9–10). This unquestioned privilege given to *Fragen* installs some beings in Heidegger's existential analytics as privileged over other, since *Dasein* is the one holding the privilege to question.

Having experience of the question, *Dasein* is exemplary in its openness to the questioning of being. Thereby *Dasein* differs from the animal, which lacks this experience. Derrida identifies a paradox in Heidegger's discussion of man and animal. *Dasein* has a world, or rather, *Dasein* is world-forming (*weltbildend*). While the stone is without world (*weltlos*), the animal is poor in world (*weltarm*). To be poor in world suggests not a complete lack of world. If the animal is deprived of world, it must be because it *can* have world, meaning that its privation (*Entbehrung*) of world is a not-having world. Still, to be poor in world is not a quantitative difference but a structural and qualitative one. It is a matter of perspective; the animal is poor in world from the point of view of man, from a non-animal world (ibid. 47–50). However, if the animal's privation of world is such from the perspective of *our* world, one can once more begin to discern the contours of a community predicated on sameness and difference.

In *Parmenides*, the animal poverty of world is discussed in relation to the open in a poem by Rainer Maria Rilke. Heidegger emphasises how Rilke's account of the open as a constant and unrestrained progression of beings is not the same as the cleared and freed open of which Heidegger himself speaks. When writing of the open in his Duino elegies, the creature plays the central role for Rilke. The creature sees a limitless open that is essentially distinct from an open thought as the essence and truth of *aletheia*. While Heidegger's open finds its openness in a questioning relation to being, Rilke's open is grasped as an endless progression of objects. According to Heidegger, there is no opening for Rilke's animal; to the extent that the animal is in the open, it cannot perceive its openness. Humans, in contrast, are bound to relations with objects and other beings and are hence in the midst of an openness which has the character of a strife between disclosure and withdrawing. This is ultimately so because humans, contrary to animals, have the *word* (Heidegger 1992, 154–60).

Having the word, humans relate to entities *as such*. This is the structural difference between humans and animals. Relating to entities, animals have some form of world, but they are also deprived of world since they do not relate to entities *as such* and in their being. The lizard on the rock in the sun relates to the rock and the sun, but not to the rock *as* a rock and not to the sun *as* sun. It relates not to the rock and the sun *as such*, "as that with regard to which, precisely, one can put questions and give replies" (Derrida 1989, 52). While not closed to the entity, the animal is closed to the opening of the entity. It lacks the capacity to question the entity due to the "subtle but decisive phenomenal structure of the 'as such'" (ibid. 55).

Derrida argues that this in-betweenness of the animal has problematic effects for Heidegger's analysis. *First*, the rift between the living and the non-living and between animal and human *Dasein* installs an anthropocentric teleology, in the sense of a hierarchisation emanating from a discourse on poverty and privation (ibid. 55–56). *Second*, the animal defies the categories of Heidegger's existential analytics. The animal is not *Dasein* since it has no access to the 'as such'. It is also not *Vorhandenheit* or *Zuhandenheit* in the human world, since it defies the thingness of the thing—having a world, albeit another, it cannot be encountered 'as such'. This means that "one no longer knows what modality of Being to reserve for the animal—for itself and for us, for the human *Dasein*" (ibid. 56).

Thus, the animal is poor in world, at least from the world that forms *our* point of view. Therefore, a delimitation of 'us' is made with the animal so that 'we' are demarcated against an other so fully other that no encounter appears possible. This means that the animal modality of being, which the deconstruction of ontology lacks tools to decide, must also affect what 'we' are. Derrida asks if the obscurity of the animal does not threaten the entire deconstruction of ontology and the existential analytics of *Sein und Zeit*. Is this not compromised, he wonders, and to which I would like to add are not 'we', "by a *thesis* on animality which presupposes—this is the irreducible and I believe dogmatic hypothesis of the thesis—that there is one thing, one domain, one homogeneous type of entity, which is called animality *in general, for which any example would do the job*" (ibid. 57, *last emphasis mine*)?

The distinction between human and non-human *Dasein* marks out a humanist teleology. To confirm the extraordinariness of the human and her relation to being, Heidegger needs to delineate against an existence that relates to the world according to another, non-human structure. Derrida maintains how this humanist teleology, albeit not in terms of progress, is "*up till now* (in Heidegger's time and situation, but this has not radically changed today) the price to be paid in the ethico-political denunciation of biologism, racism, naturalism, etc." (ibid. 56). Thus, to fend off racism, the human needs to be thought as qualitatively different from any being that can be reduced to its biology or nature. At work here is not a difference in category that can be transcended, as would be the case of a teleology of progress, but a difference in *kind*. This leaves 'us' on dark waters. If *any* example would do the job of forming this difference in *kind*, who, one wonders, would comprise the 'us' to which this difference is distinguished?

In "Heidegger's Ear: Philopolemology", the distinction between *Dasein* and non-*Dasein* proceeds from friendship and *hearing* instead of questioning. Again the animal is demarcated against human *Dasein* on the basis of a privation of world building due to a lack of language, now read as an incapacity to hear. *Dasein* can open itself to being and to the world due to its ability to hear. Hearing, Derrida says with respect to Heidegger, is constitutive of discourse. This should not be understood in a mere physical or acoustic sense. Rather, the ear is an inner ear, through which *Dasein* can open up to its potentiality for being. Hearing must be grasped through the German *Hören* and its relation to *Hörigkeit*, which designates docile listening or obedience, and *Gehörigkeit* meaning belonging. Hence, through *Hören*, *Dasein* is subjected to language and to listening while it also belongs and shares in a community (Derrida 1993, 172–74).

Heidegger's discussion on hearing unfolds around the voice of the friend that one carries with oneself. The voice of the friend differs from the *phoneme*, or the noise "perceived by an animal ear", and must instead be "an essentially understandable voice" (ibid. 174–75, at 175). Thus, there is no discourse without understanding or comprehending, and no comprehending without hearing (ibid. 173). In the voice of the friend, which opens for both obedience and belonging, there is a resonance with Rancière's disagreement.

Disagreement can come about only if heard and responded to, only when resistance assumes the form of discourse in the ear of those hearing. Now, paired with the previous discussion on the animal as a modality of being that defies both existential and categorical inscription, but that still serves to demarcate 'us'—a job for which *any entity* would do—the question of the belonging of the unheard escalates in significance.

The animal ear, which is only capable of hearing *phone*, is what irreducibly demarcates 'us'. The demarcation it marks out is also a boundary of friendship, struggle, and belonging. If any entity would do for this job of marking out, then there must be some, not necessarily animal, for whom friendship, struggle, and belonging are denied. Rancière's solution to this problem rests in the assumption of equality as given. This assumes the equality of speaking beings and, hence, beings as speaking and audible. Therefore, there is a possibility for the one not heard to make herself heard and, in so doing, move what marks the boundaries for belonging. As seen, Rancière's possibility for the subjectification of the *sans-part* depends both on being at the liberty to speak and on being at the mercy of the other's capacity and inclination to hear. Still, up to a certain point, the *sans-part* has a say in herself being heard, in contrast to Heidegger's animal. The *sans-part* can push the boundaries for audible discourse and insist on her own articulation of voice. The difference of the *sans-part* is a difference in position (part or no part), not in kind. These possibilities are denied to the animal as seen by Heidegger. And, if *any* entity would do to demarcate the 'us' able to engage the 'as such', then that entity, with Heidegger, seems fundamentally excluded from belonging, from world building, and from being. For both Rancière and Heidegger, the boundaries of belonging rest in language; in having access to the questionability of language *as such* and to being heard as speaking beings. Still, Heidegger's boundaries are far more rigid than Rancière's, to the extent that they narrow down the 'we' both with respect to its meaning and its space.

This rigidity and the degree of the exclusions emanating from the boundary of 'us' escalate in Derrida's discussion on the spirit or the spiritual in Heidegger. While this text will not follow all of Derrida's extraordinary and haunting deconstruction of spirit—of *Geist*, *geistig*, and *geistlich*—in Heidegger, I here wish to focus on how Heidegger's notion of spirit ties onto language in general and some languages in particular. First, attending to *Being and Time*, Derrida sees how 'we' always already are inscribed within a metaphysical understanding. Asking the question 'what is Being?', 'we' are already bound up with the 'is'; 'we' cannot escape proceeding from an "is" which gives the question a certain guidance from the start. "*This average and vague understanding of being is a fact*", Heidegger asserts (Heidegger 2010, 4), though this means that 'we' always already act within an understanding of being, and that 'we' are determined by this understanding to which 'we' can have a questioning relationship. Determined by the 'is', the question of being is inscribed in a horizon of metaphysics, and in an Indo-European linguistic milieu.[1] It is within and to this milieu that the question of being is raised. And as seen, what is proper to man is proper to being, so intervening in the factum of being means to intervene in the determination of the factum of man and in the determination of the factum which 'we' are. But this intervention is always an intervention within an Indo-European linguistic milieu, a milieu in which being is always already inscribed, and in which also 'we' are inscribed. Hence, 'we' are already determined within a Western metaphysics (Derrida 1982, 124–26).

Interrogating the question of the question and the privilege of questioning, Derrida identifies how a promise comes prior to any question to Heidegger. Before any question, language is already there. At the moment when any question is raised, we must already be within language. Language is there in advance; "it exceeds the question" (Derrida 1989, 129). "This advance is, before any contract, a sort of promise of originary alliance to which we must have in some sense already acquiesced, already said *yes*, given a pledge" (ibid.). Thought, Derrida writes, "is fidelity to this promise. Which means that it is only what it should be if it listens—if it both hears and obeys" (ibid. 94). Both *hörig* and *gehörig*, both docile (*fügsam*)[2] and belonging, thought gathers in questioning. The belonging is a belonging to the promise, to the "commitment of language towards language", to "this giving of language by language and to language" (ibid. 129). Again, belonging is set up in relation to language, to the fidelity pledged to the promise of language. But this promise "would also be a promise *of spirit*", Derrida underscores (ibid.).

And here 'we' are yet again narrowed down from the *Mitsein* of a human *Dasein* demarcated against the animal. It is delimited from a 'we' inscribed in Western metaphysics, to a 'we' of a highly particular historical constellation. With the spirit, which forms the promise of language giving language for thought to questioningly hear and obey, 'we' now unfold only in the languages Heidegger sees as able to grasp *spirit*. That is, 'we' unfolds in the triad set up by the Greek *pneuma*, the Latin *spiritus*, and the German *Geist*. Thereby, Heidegger relates not only some languages as more bound up to each other, to spirit, to history, and to belonging than others; he also excludes other languages, which, historically and in terms of translation, certainly must have been intertwined with both *pneuma* and *spiritus*.[3] In addition, the closure of this triangle and especially the previously seen privileging of the German and the Greek form a certain foreclosure, simply by deciding this privilege and special relation without the possibility of a Greek affirmation of this joint privilege. But there is another dissymmetry to the privileged relation between *pneuma*, *spiritus*, and *Geist*, since Heidegger maintains that none of the two first languages has any word to translate *Geist*. *Geist*, spirit, is not first of all "breath or pneumatic inspiration" but "refers *first of all* to flame" (ibid. 70–71, 99–100, at 71). Hence, "it turns out then that of the two twinned languages, Greek and German, which have in common the greatest spiritual richness, only one of them can name what they have and are in common par excellence: spirit. And to name is to offer for thinking" (ibid. 71). Hence, the demarcation—due to the fidelity to a promise of language to give language, to gathering beings out of their belonging to thinking and questioning, and of relating to entities 'as such'—that first demarcated against an animal *Dasein*, but for which any entity would do the job, now demarcates an 'us' so narrow that most of us are left out. Derrida's deconstruction has shown a privilege of the human turn into a privilege of the West, to eventually land in a privilege of the German.

Following this path of thinking, why would one, as for example Derrida continued to do, at all engage with a thinker such as Heidegger? Even if one can make the choice not to follow Heidegger all the way to the particularity of the German language, it still remains that 'we' are structured through language according to the 'as such'. Consequently, while 'we' is open to questioning—and an attentiveness to being should indeed call for a continued questioning also of the meaning of 'us'—this questionability comes only at the cost of a primary delineation, of, one could say, a temporary sameness that highly presupposes difference. As Jean-Luc Nancy puts it, "there is no universal 'we'. ... 'We' is said each time of some configuration, group, or network, however small or large" (Nancy 2000, 76). Although 'we' are transitory and changing, 'we' are always *something* (ibid.). And here lies a central problem in Heidegger's thought, at least for the purpose of thinking a community *not* predicated on sameness; the thinking of being, of questioning, and of philosophy is bound to its pledge to language. Language ensnares existence; it binds existence while also releasing it to openness and to questioning.

Now, Derrida's reading of Heidegger can of course be questioned. In his highly praising review of *Spirit*, John Sallis lifts another aspect of Heidegger—that there is a speech before the word. While Derrida himself draws on this in a long note to *Of Spirit*, Sallis suggests Derrida underplays its importance for Heidegger's thinking. Sallis further notices how Derrida throughout his writings on Heidegger "has sought to expose ... an element that would limit Heidegger's work, constraining it in this respect within that very conceptuality that in other respects it decisively interrupts" (Sallis 1993, 129). In *Of Spirit*, this limitation would be the experience of the question. In contrast, Sallis addresses Heidegger's attempts to think the question in a more originary way and notices further how the late text *What Is Called Thinking* reveals that thinking and the structure of the question as the comportment of *Dasein* "is no longer—if it ever was—to serve as the point of departure" (ibid. 128–30, at 129).

For the purpose of this book, what emerges as complicit in the priority of language is how 'we', a questionable, transitory 'we', forms a 'we' bound to be said in ontic metaphor. Only *as such* can 'we' be open to questioning, suggesting that also the swirling character of the questionable Greek *polis* is closer to the readily presentable than Heidegger might wish to admit. Complicit in the priority of language is also language itself, for there is no metalanguage that can address language from its outside. To question, 'we' must already be handed over to language, but a language that is ours and leaves us to be questioned only from *within* whatever understanding of us. If 'we', then, is already demarcated against some modality of being, it seems that 'we' cannot escape being formed on the relation between sameness and difference.

THE DIFFERENCE OF LANGUAGE

Over the course of this book, language has returned as a promise and problem of the thinking of the political community. The very relation of identity and difference as always presupposing each other was ascribed to a conceptual thinking. Formulating difference within the realm of conceptual thinking, it was suggested, means to inexorably tie difference to identity and identity to difference. As a consequence, attempts to augment identity are bound to summon new identities and, with them, new differences. Working under the spell of representative thinking, it was said, means that contestations of the limits of community will give rise to new limits, since sameness as the condition of possibility of community is left uncontested.

From the initial question up to this point, the problem of a community predicated on sameness has grown in abstraction. On the one end, the problem is one of exclusion. It is the hardships, suffering, and struggles of all those beings who fail being encompassed by whatever notion of sameness the community is founded on. On the other end, it is a problem of metaphysics, making sameness an immutable feature of the metaphysics of community. In between, the notion of a community predicated on sameness has been formulated as a problem of a power, both of sovereignty and imagination, which streamlines time and space and arranges coexistence under the order of homogeneity and calculability. Finally, it has emerged as a problem of the political, since it appears that the problem of a community predicated on sameness forms the condition of possibility of politics. In all instances, conceptual difference forms a central part to the emergence of community and the possibility of politics. Alongside being a tool for control and domination, language is also a vehicle for change and resistance, though it will only take us so far as to its next determination. We have seen two instances where language directly bars engagement and belonging: one in Heidegger's emphasis on a language that leaves being, belonging, and world building available only to those few with access to this particular language; the other in a reworking of speech as predicated on the assumption of equality, something that excludes the elevation of one language but still leaves those in the most precarious of situations without part in the presentation of the community.

Nancy finds an opening out of the deadlock between being, belonging, and world building on the one side and language on the other in a heterological meaning circulated between beings. For Nancy, the question of being is less how one engages oneself and the other. Instead, it is essentially a question of being-with. Rather than proceeding from an analysis of *Dasein* to *Mitsein*, Nancy addresses the need for existential analysis to take being-with as its point of departure. To Nancy, being is given to us as meaning. However, being does not have meaning; instead, meaning consists in its own circulation and "*we* are this circulation" of meaning (Nancy 2000, 2). Such circulation of meaning goes at once in all directions of all space-times. It involves all beings and entities, inanimate or animate, dead or alive, future or past. It also involves humans, that is, "those who expose sharing and circulation *as such* by saying 'we'" (ibid. 3).

To engage meaning means to simultaneously engage sharing. Indeed, to Nancy, *with*, or rather *being-with*, has for philosophy been secondary to *I*. For Nancy, one makes sense only in relation to another. Therefore, being one is essentially being-one-with-another. Attending to being as being-with turns presence into copresence, appearance into coappearance, and existence into coexistence. It also suggests that the singular can only be grasped in relation to the plural. Hence, being is nothing but *being singular plural*. A singular being implies already its own singularisation and differs thereby both from the individual and the particular. What is ideal to the concept of the individual is an immanent totality without any other, that is, without anyone that is not an individual. The concept of the particular, similarly, assumes the totality of what it is part. The singular, in contrast, "is primarily *each* one and, therefore, also *with* and *among* all the others" (ibid. 32). This should not be understood as an I who is its own origin and essence and exists together with others who are also their own origins and essences. At stake is instead to grasp an I formed in the continued turn to itself from the others with which it is.

In the same way as being singular plural draws a one that must always be understood as one-with-another, it poses origin as shared *at* the origin. At work here is a plurality of origins, though not in the sense of one fixed and isolated origin next to other fixed and isolated origins. On the contrary, the plural unfolds at the singular origin so that this is plural in relation to itself. Circulation of meaning signifies a "sense of stretching or spreading out [*tension*] from one origin-of-meaning to another" (ibid. 87). Meaning is circulated back and forth between singularities; what is said by me is resaid by another, so that "what comes back also becomes another origin of meaning" (ibid. 86). Here, in this circulation of meaning, the nature of the common is brought into play. The common, according to Nancy, is not a sphere of unity; on the contrary, it is established through the multiplicity of meaning, that is, the circulation and exchange of possible experiences. Therefore, meaning is commonly shared and necessarily outward reaching. What democracy then means is that multiplicities of meaning are allowed to be part of a community, which is not uniting them but brings their multitude into play, into *tension* with each other (Nancy 2010, 80–81).

Nancy renders the between, the with, as a *spacing*, although without further exploring the spatial implications of the term. Being-with requires a spacing because it demands a distance, or rather, a disparity or dissemination. Nancy describes this as a *dis-position*, which contrasts to the Position in which a stable origin would be fixed. While the Position prevents circulation of meaning, the spacing of the disposition forms an opening for such circulation. Dis-position has also bearing on the meaning of community in bringing to light being-in-common as the dis-position of the community so that the latter must be grasped as without common origin (Nancy 2000, 20–23). Thus, not only is the meaning of 'us' formed in circulation but 'we' are also conditioned by it. 'We' are always something, though however conditioned on the dispositioning and circulation inherent to being-with. If the with forms the point of departure for philosophical thinking, then with or being-with must be kept open, meaning that "we have to disidentify ourselves *from* every sort of 'we' that would be the subject of its own representation" (ibid. 71). We must not think 'us' within representational thought, but rather as the *praxis* and *ethos* of "the staging of co-appearance" and "the staging which is co-appearance" (ibid.). Hence, "We are always already there at each instant. This is not an innovation—but the stage must be reinvented; *we* must reinvent it each time, each time making our entrance anew" (ibid.).

Whereas Heidegger displaces the origin so that it lies unreachable and unidentifiable beyond the *Ereignis*, Nancy erases the difference between being and beings. This means not that he returns to objective presence, subsumed under an all-encompassing *logos*. Instead, he poses a plurality of origins thought from the singular plural of being-with. Nancy writes, "If the world does not 'have' an origin 'outside itself', if the world is its own origin and the origin 'itself', then the origin of the world occurs at each moment of the world" (ibid. 83). Each moment is already marked by the with; that is, it is marked by a meaning that it is heterogeneous at its origin. Thus, the circulation of meaning keeps meaning open and necessarily outward-reaching. The said is always from one origin to another and returns from another origin to a new origin (ibid. 87).

Still, while Nancy's being-with makes meaning irreducibly immanent and lets existence always be thought in plural, it stands in a peculiar tension with representational thinking. 'We' should not be thought as representation, he says; still 'we' are always *something*. While the dis-positioning of 'us' breaches the position and sustained identity it demands of the individual or of 'us', it also implies the overcoming and contestation of each position. Thus, politics emerges again as a matter of undoing boundaries, interrupting sameness, and dislocating, but also relocating, identity. Further, the origin that I am is transferred to the origin of another, where it is transformed into a new origin sent back to me to land as yet another origin. Each origin is disrupted and displaced in the circulation of meaning. Still, what returns is a new origin that one might hang onto for a while, until it again is interrupted through being-with. Despite the multiplicity of origins, they are not heterogeneous to themselves. Also I am always something, meaning that for Nancy self-determination, albeit transitory, remains.

Also Derrida stays within language. While he reveals Heidegger as a prisoner of language, and as ascribing a certain idealism to language and specific languages, Derrida himself would not give up on language altogether. On the contrary, to Derrida there is neither a prelinguistic ground nor any thing or space that is not already caught in text. So while rejecting the Heideggerian idealism of one idiom, Derrida embraces the inherent difference of language. As he elaborates in a dense note to *Of Spirit*, the forfeit, the *gage*, "paid to language is not a prelinguistic silence", but an "en-gage-ment in language and a responsibility to it" (Krell 1993, 31). To surrender to language means to engage in a responsibility to language by employing the latter as a respons-*ability*. Respons-ability means that one must hear, but it also unsettles the positions of the hearing self and the speaking other. This turns *Hören* away from obedience and belonging, into an issue of the very possibility to obey and belong through language. This means to grasp the very singularity of an idiom, though to do so, one must also hear its plurality and attend to "the *Auseinandersetzung* of idioms" (Derrida 1993, 215). For a hearing imbued with respons-ability, hearing must expose itself to its plural character. As Derrida puts it, "more than one ear is necessary" (ibid.).

Monolingualism of the Other (Derrida 1996) revolves around the issues of language, authority, and belonging. The text engages the plurality of hearing and of the idiom through proceeding from the insight that (a) "*I only have one language, yet it is not mine*" and the, in and between themselves, contradictory propositions (b) "*We only ever speak one language*" and (c) "*We never speak only one language*" (ibid. 2, 7). The contradictions mount up when Derrida suggests that (a) and (b) preclude any bilingualism or plurilingualism, while (c) implies plurilingualism as all there is (ibid. 21). Nevertheless, this makes

Derrida see a universal value: "Anyone should be able to say 'I only have one language (yet, but, henceforth, lastingly *à demeure*) it is not mine'" (ibid.).

To sustain this claim, Derrida draws on two figures: the figure of the colonial master and the figure of himself. Referring to his own experience as an Algerian Jew, Derrida finds himself deprived of a mother tongue, whether that is Arabic, Berber, or Hebrew. As a pupil of a French school, the supposedly maternal language was that of the Metropole, French. So if Derrida surrenders himself to language,[4] then he surrenders to the language of the other (ibid. 47). Located elsewhere, the language he speaks is a language he shall never inhabit. It is a heterogeneous language that requires its speaker to discern a trace of return to a language of his own. Yet, for the schoolboy in the colony, access to the own language(s) is barred; they are not taught in school, and perhaps even prohibited. Hence, there is no own to return to, no mother tongue waiting at the end of the path of French. Equally, French will not lead to a French of one's own, since the language of the colonisers is always situated elsewhere.

However, the master also lacks an exclusive possession over language. He has no natural hold over language but has historically appropriated and imposed it as his own. What matters to the master is "to make himself understood", that is, to make his speech act of ownership and mastery work (ibid. 23–24). Thus, on the one hand, the monolingualism of the other is the language of a sovereign originating from elsewhere. It is the language of the law and the law as language. Therefore, the experience of the monolingualism of the other is both autonomous and heteronomous. It is autonomous since the appropriation of the law requires that one speaks of the law as one's own. It is heteronomous since so is "at bottom, the essence of any law", with respect to the situation to which it is applied (ibid. 39). Further, a monolingualism of the other that assumes sovereignty as its foundation is always colonial and repressive since it seeks to reduce language to a monolanguage, to the language of the One, and thus subject language "to the hegemony of the homogeneous" (ibid. 39–40, at 40). Consequently, any liberation, emancipation, or revolution will, when declaring freedom from the master in the master's tongue, internalise and reappropriate a colonialist language.

The monolingualism of the other is not only an experience held by the colonial other. Experiencing the language of sovereignty means to experience an imposed identity assumingly inhabitable only due to a prior subjection. Thus, the monolingualism of the other will picture the other as the one subjected to an overseas coloniser but also the subject of the Metropole. Colonialism is at work also amongst the colonisers. The language we speak is always a language "that we do not *own*" (ibid. 40). It is an asymmetrical language; it exists "always for *the other*, from the other, kept by the other" (ibid.). It comes from, remains with, and returns to the other (ibid.).

On the one hand, the alienation of language inscribes colonialism at the heart of language, as the monolingualism of the other. Still, this problem is not resolved by opening up the language of the One to include all modalities of identification. As seen in the discussion on Heidegger, any attempt to say 'I' within language is already burdened with the demarcations emanating from saying 'I' 'as such'—as something questionable and contestable—precisely by being pinned down with particular meaning. Though, if there is no stable mode according to which the ego can be identified, how can I be said? To Derrida, "This *I* would have *formed* itself, then, at the site of a *situation* that cannot be found, a site always referring elsewhere, to something other, to another language, to the other in general" (ibid. 29). This is a situation of an "uncountable language" (ibid.). And here is the gist to Derrida's discussion: *there is no calculability of languages*. While a dominant language imposes its hegemonic sameness on the identity of people, the One of monolanguage is not identity (ibid. 29–30, 60). The One is always other; it is always elsewhere.

On the other hand, precisely in the imposition of a homogenising dominant language where one is subjected to the master's speech, the monolingualism of the other opens to an otherness of the other *and* of oneself. It gestures towards how anyone is homeless in the language she inhabits. Everyone is bound to a language which is always plural, for "no such thing as *a* language exists" (ibid. 65). Yet, since there is no metalanguage, since "*a* language shall always be called upon to speak about *the* language" (ibid. 69), any language is essentially homeless and inhabitable. To Derrida, this opens language as a site for sharing that one continuously must return to. Speaking is a gathering of difference; it is a promise of "the uniqueness of a language to come", a uniqueness not found in identity but in the singularity of a language "gathering together its difference to itself: in difference *with itself* [*avec soi*] rather than difference *from itself* [*d'avec soi*]" (ibid. 67–68). This is the monolanguage of the other, where the 'of' signifies less property than origin so that "language is for the other, coming from the other, *the* coming of the other" (ibid. 68).

Whereas Derrida maintains that Heidegger's abandonment of ontology never becomes an invitation to multiplicity (Derrida in Lacoue-Labarthe and Nancy 1997, 85), monolingualism of the other forms precisely a yes said to alterity and otherness. This is what 'we' share in: in experiencing that I only have one language, yet (and therefore) this language is not mine. Thus, our sameness consists in our common otherness, between us and with respect to ourselves, which is accessed through the heterology of language. This is a precarious resolution to the problem posed by Heidegger, for it draws precisely on the experience of subjection to sovereignty and to colonial dominance. Inside language there is a terror "soft, discreet, or glaring" (ibid. 23), but also a promise—a promise of the impossible possibility of all speech, a promise without determinable content, a promise of "language to summon the heterological opening that permits it to speak of something else and to address itself to the other" (ibid. 69).

The opening of language would invite us to follow the trace of language and to be open to the coming of the other. To better grasp the advantages or disadvantages of this for thinking a community not predicated on sameness, the following section turns to the relation between the calculable and the incalculable as engaged by Nancy and Derrida in order to nuance the problem of the incalculable.

CALCULATING WITH INCALCULABILITY?

What was here called a predicament of situation formulates a notion of sameness out of the shared problem of dwelling in a measureless place, and in letting a manifold world unfold from this place, in *the face of the authority of a count*. While this delimitates 'us' according to the reach of the effects of calculable thinking, it keeps 'us' open with respect to the differences of situation. To the extent that we are touched by a rationality of calculability and sameness, we all share in being victims to a rationalism that cuts off our embedded relation to the world, be it by force or forgetting. Chapter 1 engaged the stubborn problem of a democratic people predicated on its incalculability and a democracy that relies on counting. It was said that the condition of possibility of demo*cracy*, namely calculability, formed the condition of impossibility of the democratic people. Chapter 2 deepened this problem by revealing a notion of *demos* that relied on an indifference to difference which assumed no positions as grounded, but simultaneously installed different positions, hierarchies, and mastery as necessary for democracy. This meant that the lacking legitimacy of any democratic mastery would form a ground for both a calculating power and an uncountable *demos*, though only whilst precluding the persistence of the latter. Although the predicament of situation might exceed democratic thinking, it could be grasped as the problem of being a democratic people in the face of democracy.

Thus, we stand again before the aporia of democracy, which is also a problem of incalculability. For, on the one hand, the democratic people require a notion of freedom that is not established against a measure, in other words, that is incalculable.[5] On the other hand, a demo*cratic* people imply some form of mastery and thus invoke the problem of who counts. This problematic is explored by Derrida in his reading of Nancy's *The Experience of Freedom*. There, he suggests the aporia of democracy as well as of the political to emanate from an undecidability between the calculable and the incalculable.

This undecidability stems from the relation between freedom and equality. I will here follow Derrida's discussion of Nancy's shared freedom, since the latter's notion of freedom draws on similar sources as my characterisation of place. This helps nuance the discussion on the predicament of situation and the problem of calculability and incalculability it describes.

Nancy draws freedom as shared, as "what spaces and singularizes" (Nancy 1993, 68) and what "throws the subject into the space of the sharing of being" (ibid. 70–71). The task of thinking is to engage freedom before or beyond any established meaning, that is, a freedom that is incommensurable and incalculable and does not take the form of property (ibid. 70, 78–79). Its problem, however, lies in sharing freedom in an "equitable, equal, and measured fashion" (Derrida 2005, 47).

There is nothing new to the antinomy between an incalculable freedom and an equality that forms a calculable measure; on the contrary, it has long been recognised to lie at the heart of democracy, Derrida reminds. Indeed, these two laws of democracy are not always in opposition; when everyone is equally free, "equality becomes an integral part of freedom and is thus no longer calculable" (ibid. 49). Thus, equality forms the unconditional condition of freedom as well as of sharing. However, Derrida notices how Aristotle's two notions of equality—equality of number and equality of worth, or *logos*—form also a prior delimitation to both equality and freedom, since they only count those beings who are already equally free, and therefore also "incalculably, unconditionally equal in their freedom" (ibid.).

Let me note in passing the similarity between the two accounts of equality above. Both equality of number and equality of worth or *logos* rest on an assumption of sameness, either of a measure for counting or a measure for what counts as *logos*. Consequently, even if equality before the law suggests all subjects as equally free, it does not imply a freedom without measure, since the determination of equality through sameness has already saturated the meaning of freedom with identity.

Nancy's suggestion that freedom means "to measure oneself against the nothing" (Nancy 1993, 70–71) is opposed by Derrida, since it reintroduces 'the self' and thereby the problem of *ipseity*. That is, it installs *ipse*-identity so that 'the self' is put "in the position of taking the measure of existence" (ibid. 71). Equality is now reintroduced in the form of a technical measure of equality that will neutralise all aspects that bar access to the immeasurable. Hence, the immeasurable is given access to through a calculable measure. Here, Derrida identifies three incompatible necessities.

First, the technical measure of equality is not a mere "mishap" of the immeasurable but its very chance. It is what gives access to the incalculable. Further, political, juridical, and ethical responsibility presupposes such a technique and is rendered some meaning only through the technical measure of equality. *Second*, as a chance to get access to the immeasurable, a technical measure of equality is also a threat, since its calculation destroys the immeasurable it should give access to. And *third*, since there will be no common rule to regulate the calculation of the incalculable, the question is where it ends. At citizenship, at the borders of the nation-state, with all non-human living beings, or with the nonliving? When not understood as the freedom of an *ego*, freedom could be extended to everything appearing in the open. The problem for Derrida is the problem of demo*cracy*, for if freedom is extended to anything emerging in the open, including the multiplicity of the unconscious, how, then, "on what unit of measure and on what technique", should democracy calculate (Derrida 2005, 53–55, at 55)?

As concerns the question of measure-taking as the method to get access to the immeasurable, its problem is that it installs rationality as the means to overcome a certain rationale. Therefore, the problem of determining who is encompassed by the incalculable becomes delineated beforehand, at least as pertains to the question of who can engage a freedom that involves to *measure oneself* against nothing. Since both what needs to be transcended and the tools of transcendence are vested in a rationality of calculability, only the ones upholding such a rationale can engage in its overcoming. The question that Derrida directs to Heidegger can therefore also be directed to Nancy—if the question of being, which Nancy reads as the question of the sharing of freedom, envisages Nietzsche's superman, or if we in this question find the "last sleeping shudder of the superior man" (Derrida 1982, 136)?

Derrida's own engagement with the calculable and the incalculable carries resemblances, though he is less inclined to set up the problem in terms of an opposition between the measurable and the immeasurable. On the contrary, he is troubled by how he sees Nancy posit the immeasurable as a *negation* of measure. Instead, he wonders if the immeasurable must not be seen as *heterogeneous* to all measure (Derrida 2005, 57). To Derrida, the aporia of the calculable and the incalculable is therefore not one of measure, since this inscribes the two within the same logic, but draws instead together two heterogeneous rationalities.

Before proceeding to discuss these heterogeneous rationales, let us recall Derrida's problematisation of Heidegger's distinction between human and non-human *Dasein*. This unfolded around the problem of an understanding of *Dasein* conditioned on the possibility to question. Following Heidegger's path, Derrida saw this eventually lead up to the conclusion that only those with access to language in general and some languages in particular, and especially German, held the ability to question things 'as such' and thus to be close to being and belonging to 'us'. Derrida further asserted that the distinction between human and non-human was a price to pay for *anyone* seeking to denounce racism. Although this installs both hierarchies and impoverishment, this is a necessary evil to valuate the human as incapable of being inscribed in terms of race, biology, or nature. Indeed, Derrida asserted that his intention was not to criticise this distinction (Derrida 1989, 56). His problem was instead how *Dasein* is separated from another modality of being by relating to the world through questionability, and thus as presumed to have access to the world 'as such' and only from the few languages that allows for this relation.

Consequently, while Heidegger has a particular way of denouncing racism that leads him to embrace what could be called a linguistic form of nationalism, resorting to a humanist teleology is nothing he is alone in being guilty of when fending off biologist and racist thinking. Instead, "*up till now*" anyone taking up the cause against racism is bound to formulate a humanist theology. Though, as John Sallis observes, Derrida's analysis of this logic is a logic in itself, which reintroduces the measure of man in the face of racism and biologism (Sallis 1993, 125). Hence, there is another way to read the problem than simply saying that humans are more than nature, race, or biology.

If read in terms of calculability and incalculability, racism and biologism would be examples of calculable thinking, while existential *Dasein* orients herself to her world through incalculability. If these are opposed to each other, it suggests that incalculability forms a logic that takes the overcoming of calculability as its point of departure. Incalculability is then a rationale of its own, which must take the calculable into account so as to transcend it. This describes the same problem as identified with respect to the immeasurable elaborated by Nancy. If 'we' humans unfold in the strife between the calculable and the incalculable, then the animal, lacking the word and thus the ability both to count and to question the count, is fully different to the human. This difference is a difference of having and not having *logos*, since a human *Dasein* demands a *logos* that can take the calculable into account so as to reject it. Although this opens a community of incalculability, it is only deployed by putting to work a *logos* that has the ability both to calculate and discard calculation. Again 'we' is carved out through a notion of sameness that presupposes difference, that is, the sameness of man and his ability to transcend his own readily available presence. The sameness at work is a sameness that invites the incalculable, but only by keeping with reason and must thus describe a superior man. This, it seems to me, must be the meaning of the cost of a humanist teleology; that it is through a recourse to *logos*, albeit not as *ratio*, that the calculability of racism, biologism, and naturalism can be held at arm's length.

Now, how do the two heterogeneous rationales of the calculable and the incalculable, which Derrida posits, relate to the above? What Sallis identifies as Derrida's "logic" is a logic that emanates from the aporia between calculable and incalculable reason. To Derrida, calculable reason unfolds within a horizon of teleological unity that orders, orients, and makes possible historicity within a program of calculability. As doing so, it makes impossible the unfolding of history since anything unforeseeable and incalculable is already annulled by being grasped according to the particular teleological order. In chapter 1, the temporality of cosmopolitan citizenship emerged as an example of this (Derrida 2005, 128, 142–43).

Referring to Husserl, Derrida describes this as an objectivist naïvité or irrationalism by which reason throws reason into crisis. The crisis of reason is produced in the production of ideal objects that "by their iterability … cover over or consign to forgetting their historical and subjective origin" (ibid. 127). It consists in the failure, insufficiency, and irrationalism of reason. This should not cause us to give up on reason, Derrida maintains. Instead, reason must be saved from the confines of calculability. Reason must be reasoned with so as to be exposed to how rationality is not limited to the calculable (ibid. 126–33).

Saving reason means to redeem an event displaced by teleology and bring it back to its eventfulness in which it is only possible as im-possible. This is a "matter of thinking reason, of thinking the coming of its future, of its to-come, and of its becoming" as *other*, "as the absolute exception or singularity of an alterity that is not reappropriable by the ipseity of a sovereign power and a calculable knowledge" (ibid. 148). The event must be thought as the unconditionality of the incalculable. It must be thought as absolute singularity. Without this singularity of the incalculable, "nothing *other* and thus *nothing*, arrives or happens" (ibid.). The incalculable event is neither produced by constative nor by performative acts. A constative act will declare the event and neutralise it from the start, strip it of all incalculability. Also the performative will annul the event by seeking to secure it through its own calculable mastery. Instead, an incalculable event must "affect a passivity" and "touch an exposed vulnerability" (ibid. 152).

The difference between calculable and incalculable reason can be grasped as the difference between grounding and running aground. Grounding consists in the intentional and autonomous manner by which calculable reason saves itself from the eventfulness of the event. Running aground connotes instead "the moment when a ship, touching bottom, gets accidentally immobilized" (ibid. 122). The latter seems to be played out on the limit between the calculable and the incalculable, where ground-giving reason is exposed to the abyssal character of the ground. Running aground forms that tremor by which that which is accounted for is disclosed to an abyss, an *Abgrund*. It is an unforeseen and uncalculated event, happening when the ship, sailing freely, suddenly runs aground so that its rationality becomes aware of its own limits. Grounding, on the other hand, connotes an intentional and calculated responsibility for touching bottom. Though as grounding, reason itself would withdraw from any *Abgrund*, a grounding reason would seek to eschew any tremors that would shake it in its ground and open this to its own particularity.

However, Derrida never gives up rationalism altogether, for deconstruction is also a rationalism. It is indeed an unconditional rationalism that unconditionally criticises all conditionalities (ibid. 142). Hence, the condition of possibility for the potential of deconstruction is its unconditionality, while the condition of possibility for deconstruction to be exercised is conditionality. To Derrida, the unconditional and the conditional are inseparable, as are the calculable and the incalculable. For example, justice must always be heterogeneous to law. There is an unconditional incalculability of justice that always exceeds a law conditioned on norms, rules, and calculability. To apply a norm or rule means to impose a calculable order on the heterogeneity of lived experience. Still, Derrida holds that there "could be no justice without the appeal to juridical determinations and to the force of law" (ibid. 150). On the other hand, "there can be no becoming, no transformation, history or perfectibility of law without an appeal to a justice that will nonetheless always exceed it" (ibid.). To think of justice, it is necessary to think together the heterogeneity and inseparability of the calculable and the incalculable.

So for Derrida there are two rationalities: on the one side, a "reasoned exigency of calculation or conditionality", on the other side "the intransigent, nonnegotiable, exigency of unconditional incalculability" (ibid.). This is what the responsibility of reason would need to take into account. If these two exigencies of both calculable and incalculable reason should be situated, they would be so, Derrida says, in the "autoimmune aporia" of the transaction between the calculable and incalculable. Autoimmunity draws the aporia as self-consuming, though not in the sense of an eventual disappearance of the aporia. What is consumed is instead any unified measure to engage the impossible transaction between the calculable and the incalculable. Thus, there is no technical measure of equality to guide the relation between the calculable and the incalculable. Instead, each transaction between the two must invent its own law in order to be open to the event to come (ibid. 150–51).

In many ways this aligns with how I describe a predicament of situation that requires us to cast off any calculated positioning and strive for an incalculable place that can never be found through any guided route. However, for Derrida, in order to let the incalculable arrive, *you need to have access to reason*. If "reasonable" means the "reasoned and considered wager" between the calculable and the incalculable, this still gives a priority to an inside, to an 'us' that lets the 'other' arrive (ibid. 151). In a sense, then, Rancière is right to criticise Derrida for establishing an ethical community, though not by his disqualification of brotherhood and substitutability. Instead, while those of us waging the impossible transaction between the calculable and incalculable would be highly political (also by Rancière's understanding of the word) and augment the meaning of the community and the boundaries of its law, the other, the one to come, appears only as political by causing those parts of the community to engage in the wager between the calculable and the incalculable.

If the coming of the other is a wholly incalculable event, then the other is the one who precisely does not keep within reason. As fully other, the unconditional incalculability of the other is neither political nor is it part of the community in and of itself; the other is only part of a community to come. Its emergence is political to the extent that it is responded to in the irreconcilable transaction between the calculable and the incalculable. It is part of the formation of the community only if 'we' let it arrive through expanding the conditions for 'our' norms. Peculiarly, then, Derrida's justice as deconstruction appears haunted by a similar problem as Matthew Craven sees belonging to human rights; the ones at their normative core, the ones in need of protection, are excluded from the discourse and not claimants in their own case (Craven 2006, 84). If wholly other, though affecting a non-formalised transaction between the calculable and incalculable, the inclusion of the other always relies on those possessing calculable reason.[6]

The monolingualism of the other would go against this conclusion and suggest that the other is always with us, whether we calculate or not. Still, it draws its energies on sovereignty, since otherness emerges from the language of a law always different to its circumstances. Here lies the opening for a community to come: in realising that we all—sovereign and subject alike—are bound to a language that is not our own. To engage it means to take into account the incalculability of a language that cannot be affected by our calculable thinking, since this would neutralise its eventfulness. Still, while keeping with sovereignty, the community vested in the monolingualism of the other must continue to play out a transaction between calculability and incalculability.

Derrida's aim is never to throw sovereignty overboard, and certainly not with reason to follow. On the contrary, he seeks to keep within reason, so as to reason with sovereignty. For Derrida, sovereignty helps not only to show how anyone has to surrender to a language that is not their own, but in addition how nation-state sovereignty can be "an indispensable bulwark" against international powers that would imprint their rationalisation of the world "under the cover of liberalism or universalism" (Derrida 2005, 158). Still, Derrida asserts the necessity of deconstructing nation-state sovereignty, its principle of indivisibility, and its right to the exception. He even sees this as already under way through the perfectibility of human rights. However, when resorting to nation-state sovereignty as a bastion against, for example, a limitless liberalism, he underplays how they both are different expressions of sovereignty as a symbolic form that on the one hand is boundless and on the other fortifies territorial boundaries; both cases rely on a space of authority as a space of sameness.

So how does the predicament of situation fare when confronted with the calculable/incalculable copula? As long as the predicament of situation involves attending to how things are made sense of through a framework of sameness, it is likely to invite the same revalorisation of humanity and its exclusions as was seen to emerge not only for Heidegger but also for Derrida. Thus, if it suggests that calculability should be assessed in order to be transcended, it would be difficult to claim a sharp dividing line between the predicament of situation and Derrida's autoimmune aporia of the impossible transaction between calculable and incalculable rationality. Most likely, the predicament of situation is, for the many of us not only exposed to but also participating in an imposed calculability, a question of reasoning with ourselves on how to embrace the incalculability of existence. If Heidegger is correct and *Dasein* is bound to be a sense-maker, then humanity would form a certain modification of the situation that renders the predicament of situation as the impossible task of undoing the power of signification. Certainly, this task is conducive to a moral orientation towards the openness of place. Still, the attempt to release humanity from calculability and sameness would hinge not only on the incalculability of existence, but also on there being a way of reasoning with sense-making. This leaves 'us' again standing before modalities of existence different from 'us', where 'we' can just hope that the lines of separation between sense and non-sense overlap with our ethical self-projection.

If the predicament of situation would be understood only as a moral duty to transcend sameness and calculability, then any moral commitments would be placed in the realm of *logos*. Derrida confesses to as much when he asserts how a technical measure of equality, that is, a measure that can be calculated upon, is something that juridical, ethical, and political responsibility requires. It is only such a measure that gives access to the immeasurable, he maintains. However, this book suggested that beings take part in the formation of the community through an incalculable place. To approach place through a measure would mean to encroach on its difference and bring the presencing of the community to a halt. Moral commitments in an incalculable community can therefore not unfold from *logos*. Instead, if place emerges with dwellers, the sole moral principle of an incalculable community must be to let beings dwell. Morally, an incalculable community thus rests in *the reciprocity of letting beings be as dwellers and in place*.

Derrida's distinction between law and justice invites a similar conclusion whilst also revealing the problem of a morality that hinges on *logos*. It has been alluded to how a community not predicated on sameness escalates the very question of law, since legal collectives were suggested to require both selfhood and sameness. To both Lindahl and Derrida, the perfectability of human rights involves the continued interrogation of the limits of life (Derrida 2005, 151; Lindahl 2014, 247–49).[7] If human rights should not fall into a universally claimed yet irreducibly limited ordering of behaviour, places, times, and subjectivities, it needs to engage a responsiveness to what is *strange* to it. This book showed how the river conjured a place that was *there* yet always shifting. While there before us, place is always strange. To make way for its universalist claims, human rights would need to involve something of a "right to place" that takes into account the strange as already there. Certainly, to put right together with place means to risk the very meaning of place as here elaborated, since responsiveness, even if serving to augment the existing order, will always neutralise the strange and level "down the extraordinary to a variation of the ordinary" (Lindahl 2014, 266). Thus, an incalculable community does not easily lend itself to juridical and normative thinking. The predicament of situation can serve as guidance for normative discussions, but the difference of place will always reveal their limits.

Still, that the predicament of situation and place as difference does best to refrain from reasoning with the incalculable means not that they are without normative and political import. Quite the contrary, the above outlined notions of a sameness and difference that avoid representation radically address the problems of calculable communities. What, then, are the implications of the predicament of situation and place without difference for the here discussed problems of democracy, citizenship, and sovereignty? And how do these notions help understand an incalculable community? It is now time to address these questions.

NOTES

1. See further *Identity and Difference* and Heidegger's concluding question of whether it is the case that the structure of Western languages makes them onto-theological to their nature (Heidegger 1969, 73; Stambaugh 1969, 18).

2. *Fügsam* must be related to the *füge*, to a dike standing in the jointedness of a presencing and absencing. In being *fügsam* to what it has to think, thought questions into the displacements of jointures, brings jointedness forth and with it the finitude of existence (Heidegger 2002b, 175).

3. Derrida mentions the Hebrew *ruah* (Derrida 1989, 100–101).

4. "For, I confess, I always surrender myself to language" (Derrida 1996, 47).

5. This means that the classic notion of freedom, of an autonomous subject able to say "I can", must be abandoned, since this relies on the ability to count oneself as a whole, cohesive self.

6. It is also somewhat curious how Derrida addresses the urge for human rights to expand its reach also to animals and realise the heterogeneity of animality, although

the briefly summoned reasons for animals as part of a universal order of rights are articulated through capacities usually ascribed to humans. Derrida suggests that none of the "conventionally accepted limits" between humans and animals resist rational deconstruction, be they language, culture, social symbolic networks, or technicity (Derrida 2005, 151). Elsewhere, his example of a communicative non-human being is his own cat. With respect to this, Nick Land observes that both Derrida and Heidegger neglect Georg Trakl's poem on rats, despite that the question of the animal in *Of Spirit* at large unfolds from Trakl's poetry. Maybe, Land says, the reason for this neglect rests in that "difference becomes unacceptable when it moves fast and unpredictably, hissing at humanity through plague-smeared teeth" (Land 1993, 50).

7. Lindahl discusses a notion of human rights vested in the responsiveness to the *a*legal. Since the legal and the illegal both mark the meaning and reach of the same order, they imply and sustain each other. By contrast, the alegal confronts the foundation of this order. The alegal reveals the limits of the complex of the legal and illegal and confronts its claim to order a legal collective according to an agreement on the meaning of joint action. The alegal also confronts the distribution of places and subjectivities within the legal collective and challenges what is claimed as the right time and place to engage in legal behaviour. For human rights to be able to handle the disruption of alegality, Lindahl argues, it must acknowledge not only how legal boundaries are including while also being excluding, but how they also "always exclude by including" (Lindahl 2014, 267). For Lindahl, humanity consists in "anonymous forms of sociality and socialization which precede the closure whereby a legal collective identifies itself as a limited whole" (ibid. 247). By consequence, what is strange to the legal collective is human.

Conclusion

This book has circled around the problem of assuming a notion of sameness that presupposes difference as the condition of possibility for the community. It has argued that as long as the community is envisioned as predicated on a notion of sameness that presupposes difference, any attempts to enhance the meaning of coexistence will continue to give rise to externalities and separations between the same and the different. This means that we in our midst will have entrenched inequalities between subjects endowed with rights, liberties, and the entitlement to raise their own political voice and persons whose movement, visibility, and existence is conditioned on the benevolence of rights holders, as is their possibility to claim their human rights. One aim of this book was to engage this problem critically, through studying the relation between sameness, difference, and identity in Western thinking of the community. This was done to enhance the understanding of how sameness pervades political imagination and what is implicated in attempts to expand the meaning and reach of the community. A second aim of the book was hermeneutical and normative. This sought to grasp an understanding of a community *not* predicated on sameness, through developing a notion of sameness that does not presuppose difference and a notion of difference that does not refer to an underlying assumption of identity.

Now it is time to see what has been achieved with respect to these matters. In the first section of what follows, an account is given of the conclusions made in the different readings of Western political thinking of the community. Here we will see the problem of community predicated on sameness transpire to our conceptions of politics, authority, language, time, and space. The second section discusses a notion of sameness that does not presuppose difference as a 'predicament of situation' and its implications for the outlined problems. How might sameness as a predicament of situation relax some of the problems emanating from a community predicated on sameness? Finally, the last section dwells on the notion of place as a difference without identity, while seeking to draw some larger ethical and political conclusions from this and from the question of the political community approached through an incalculable place.

I

The first chapter studied the relation between a community predicated on a notion of sameness premised on difference and how this meant that being part of the community came to be understood as a question of membership. In the modern state, it was said, such membership took the form of citizenship. This problem was seen to emerge with the paradoxical legitimacy of the democratic people. If the democratic people should emerge from the laws stipulating their constitution, while at the same time being the founder of these laws, then we are confronted by a chicken and egg problem: on the one hand, there must be a democratic people to found democratic laws; on the other hand, there must be democratic laws to found a democratic people. With democratic legitimacy another problem emerged. Democratic legitimacy requires that anyone subjected to a decision should have a say in the process through which that decision is established. When communal belonging is understood in terms of membership, decisions made concerning the boundaries of the community are exposed to a lack of legitimacy since they affect members and non-members alike. This implies that the borders of the community suffer from a lack of democratic justification. The reason these problems at all emerge, the book argued, was because of the assumption that communities require some form of unity as their condition of possibility.

The persistence of this assumption was seen in Näsström's assertion of, on the one hand, the incalculability of the democratic people as the manifold voices and opinions that vouch for a democratic community and, on the other hand, a democratic authority that counts every voice as equal. That, however, forms a problem because it involves a decision on who at all counts as countable. The question of the political community is thereby turned into the question of who counts. Through time, this has been answered differently. Should free white men, women, children, animals, or those measuring up to the meaning of a rational subject be seen as subjects of democracy? Deciding on who counts is a problem of delineation and of membership in the community that eventually rests on sameness. To count equally, there needs to be a minimal measure of resemblance that lets a being qualify as countable.

As a redemption to the problem of the legitimacy of the people, democratic theory saw the cultural nation step in to provide a source for unity.

Still, the cultural nation was also flawed for this purpose. In practice, cultures are not distinct and separated but bleed into each other. They form overlapping hybrids of a vast variety of influences. If common cultures should be conjured, they must be created, but that makes culture the outcome of politics instead of its ground. In search for the myth of a common culture, cultural nationalism is bound to take recourse to ethnicity and the myth of a common ancestry.

Citizenship theory in different forms has sought to undo the tight relation between the nation and the political community. However, when assuming sameness as a condition of possibility for the community, their attempts to reduce exclusions became burdened with new externalities. This was not only seen in cosmopolitan or multicultural citizenship theories that either assume a universal moral or a secured group identity as the grounds for political belonging. More flexible understandings of the meaning of groups and locally iterated versions of a cosmopolitan moral were also marred by the problem of laying the responsibility to augment the reach of communal belonging on those *inside* the group. The ones formally excluded had to rely on the benevolence and hospitality of those formally included. Even a differentiated justice that evaded positing one image of the subject of rights failed to escape turning rights adjudication into a question of subsumption under a category. Multiculturalists would thereby be vulnerable to the same critique as they forward to cosmopolitans—their unattentiveness to the variety of identities within groups.

Further, due to its reliance identity, citizenship drew inclusion in the community as premised on assimilation rather than integration. Whereas integration should grant the selfhood expressed in *ipse*-identity, a slippage into identity as resemblance expressed through *idem*-identity was continuously made. Thereby, the ideal of integration turned into an appeal for assimilation.

Three lessons are especially important to draw from the discussion on citizenship theory. First, even when a person's identity was seen as multiple and overlapping, it was points of identification that laid the grounds for a common community. It was suggested that the reason we engage morally with each other rests in us seeing something of ourselves in the other. Consequently, when no identity can be seen, no moral commitment can be made. Whatever is just too different, whatever is incalculable to us, and whatever we cannot reconcile with what we call our own cannot be part. Thus, identification means that sameness pairs up with difference, leaving us hanging with exclusions not easily breached. Second, identity as selfhood, or *ipse*-identity, which was supposed to guarantee the integration of the individual political subject without encroaching on her cultural identity, was seen to be instilled with the particular qualitative features of *idem*-identity through being located in a cosmopolitan, liberal morality and understanding of political subjectivity.

This rendered also the liberal embracement of integration as the predominant mode for inclusion as assimilation. Third, many of the citizenship theorists here studied assume a linear teleology invested with the hope that problems of exclusion will be resolved over time. Seeing historical development as something that unfolds within a particular framework means to neglect any actions outside a particular course for political engagement. This forms a historical trajectory vested in sameness. According to this trajectory, citizenship in the nation-state has formed the way forward for political communities, drawing the Western state as an ideal to reach for all other states as long as they conform to the standards of a good state with good citizens. This, however, means to impose one by no means flawless solution for political organisation on other forms for political coexistence. It also means to create hierarchies and separating the same, proper states from different, problematic understandings of the community. Thus, citizenship was seen to itself emerge as an *idem*-identity that structures an inside and outside of international society. Hence, if trying to avoid externalities to the community, one needs to be at guard against a notion of linear time that draws the community according to one historical trajectory.

The second chapter moved on to discuss the relation between boundaries and sameness. First, it discussed attempts to make the paradox of politics a productive site for contesting the prevalent meaning of democracy. It saw how agonist political theorists, on the one hand, vigorously opposed any attempts to totalise the meaning of community. On the other hand, agonist political theory was seen to rest on the presence of limits as that which calls for political action. Politics is about contestation, but that means that politics also has to maintain an assumption of sameness at the heart of the community. If to be contested, augmented, and rewritten, the community must be presented as unified and bounded. Boundaries can only be redrawn if they first are in place. With Rancière, the ones not formally part of the community emerged as the protagonists of politics and as the ones with the capacity to shift the very meaning of the community. Still, some limitations to his theory of an assumed equality remain firmly in place. The first rests in how Rancière offers no possibility to grasp any lasting community of equals; if politics should remain possible, a wrong in the distribution of the sensible, that is, how the imagination of the community is ordered, needs to be able to be identified. A community of equals would therefore not be a political but an ethical community, according to Rancière. The second limitation is due to Rancière's notion of democracy. An egalitarian community, a *demos*, takes form in the moment of disagreement. Disagreement consists in the opposition between two different logics, a logic of the democratic 'police' community, which calculates on inclusion in a particular way, and an incalculable logic of the egalitarian *demos* that expresses themselves as equal with anyone and everyone. The calculable and incalculable meet in an anarchical principle at the heart of the community, which states that there is no ground for why some people should rule over others. While forming the ground for political community, that is, a community questioning the foundations for how it is ordered, this principle forms also the ground for political power—the ground for authority. Hence, Rancière's egalitarian community rests on precisely the same principle as does a rule bound to install hierarchies and inequalities. Therefore, the commitment Rancière pays to democracy draws his assumption of equality as paired with inequality and invites sameness as paired with difference, since the egalitarian *demos* will eventually be consumed in a new count.

Further, the enactment of the assumption of everyone's equality with anyone rested in articulating oneself as a speaking being. This also involved being heard as a being that articulates *logos*, or voice, rather than as a being that articulates *phone*, or mere noise. This was seen to install republican liberty as the condition of possibility of Rancière's axiom of politics—equality. Republican liberty suggests that freedom is limited by the knowledge of us living on the goodwill of others. Thus, freedom must be understood as an absence of dependence of others. Rancière's *sans-part* is clearly dependent on those who have part in the community hearing the claims of those without part. In addition, it appears that those most vulnerable and in most need of the protection of communal belonging go unnoticed by Rancière, since they might not be in a position where they are free to enact their assumed equality without risking their own lives or the lives of their families. Learning from these lessons, it was suggested that an incalculable community would have to find a way to grasp beings as part of the community without taking recourse to *logos*, speech, or public presence.

In addition, this book found also spatiality to play a crucial role in upholding the sameness of the community. A bordered, continuous territory formed not only a backdrop to a bounded community; territory formed both the precondition and limit of state sovereignty. Arguing that the contemporary form of sovereignty has turned the image of a peaceful domestic order surrounded by an anarchical international inside out, Bartelson showed that while authority claims today are boundless, they simultaneously retain cultural difference as managed within fortified boundaries. Drawing on Lefebvre, I suggested that such an authority was sustained by a space of sameness and calculation. Lefebvre's notion of abstract space holds homogeneity, fragmentation, and hierarchies through its assumption of sameness and calculability. Although boundless, such space relies on a notion of sameness premised on difference since its extension assumes *one* understanding of space to be possible. By consequence, differences and hierarchies within that space emerge according to the same principle of differentiation. Hence, a particular account of space also sustains the assumption of sameness as a condition of possibility for the community.

Attempts to release space from the spatiality of territory or a space of sameness and calculation were seen to be developed only with reference to representational thinking and identity. While such attempts draw on a space exceeding meaning, they were formed with the explicit attempt to rearticulate meaning. Hence, the spatiality unfolding would, on the one hand, re-inscribe the political community as predicated on sameness and difference and, on the other hand, be confronted with the same problem as were the agonists—who is at liberty to speak, act, or be in place? Thus, rather than forming the locus of political community, the spatialities deviating from territoriality emerged as mere aberrations to a spatiality vested in calculability.

By contrast, this book sought to grasp an understanding of the meaning of an incalculable community—of a community *not* predicated on a notion of sameness that presupposes difference—through exploring the notion of an *incalculable space*. It thereby sought to approach the question of the political community through engaging the spatiality of the community to unveil another understanding of communal belonging. Especially two figures were used for this—Heidegger's notion of dwelling and his reading of Hölderlin's hymnal poetry on the river Ister.

Attending to the Ister revealed how the activity of the river determined the human abode through vesting it both with locality and journeying. This journeying consisted in the encounter between the self and the other, but also with the otherness of oneself. Whereas Heidegger saw the journeying unfold between two historical nodes—the German and the pre-Socratic Greek people—I maintained that the journey could be released from this particular historical setting to instead take place locally between *any* beings. Any beings, it was suggested, are able to engage in the encounter without appropriation that forms the very possibility of journeying.

Further, I argued that with the river, the place of the community must be understood as always different in relation to itself so that what is here is never merely so but is also already elsewhere. What is here has vanished but is also yet to come. As vanishing, what is here points to how entities and the world they make up come to pass, how they transform and move on so that their vanishing provides for what is coming. The river does not run aground in a temporary stabilisation of meaning through a sign or representation. It has already swept with it all articulations of meaning so that its orientation is always towards what is yet to come. Therefore, the river conjures a place before language, a place ahead of time and ahead of determination. Hence, what is to come with the river is a place that circumvents meaning and is before both sense and calculation. Both here and elsewhere, river place signifies less a process of reidentification—which is a response to a circulation of meaning and thus comes *after*—than it describes openness towards the yet undisclosed.

The figure of dwelling showed how beings gather a place through dwelling in it. Whereas Heidegger ascribed the ability to dwell and to be world building to the human, I suggested that anyone can dwell. Dwelling is carried out in the intimate relations with the world, where beings withdraw from calculability and representation and experience the world first of all as felt. In these intimate places, beings form the world through pressing upon it with their bodies. This could be seen to attribute an identity to place, but since a locale, an *Ort*, is marked by the flow of the river, I found the opposite conclusion to be correct—that *place describes a difference without identity*.

Place is thereby released from rootedness and delineations. It is neither a backdrop or container of the people and its identity, nor does it form the soil for the historical unfolding of a community. What is peculiar to the river is not only that it flows and continuously names another 'here'. What is exceptional to the river is that it is *there* in an always differing presence. The exceptionality of being there yet never the same is what makes place a crucial element to engage when intimating an incalculable community. Determining the place of the community, the activity of the river draws also the community and its site as ephemeral and in passing.

The final discussion on Heidegger, Nancy, and Derrida and their engagement with reason further elucidated the problem of speech, *logos*, and calculability. When Heidegger turned to speech and thinking to counter a framework of modern calculability, he eventually ended up conditioning existence on the ability to question whatever is encountered in the world. The problem was that this ability separated human *Dasein* from a non-human *Dasein* and formed a privilege of the few speaking a language conducive to such questioning. Consequently, only a limited few could be seen to fully belong to humanity. Nancy and Derrida were seen to open up the meaning of language to give room for difference *within* language. Still, when engaging the relation between the calculable and incalculable, they ran into problems. To secure an incalculable freedom, Nancy installed the measure of equality. While this should ascertain that one measures oneself against nothing, it also kept both the measure and the self in place, together with a community that changed in a continued circulation of meaning but which at each moment always was *something*. Derrida, in turn, addressed the problem of a grounding reason that neutralises the event and anything to come by posing this within its own rationality. To him the impossible wager between a calculable logic and a heterogeneous, incalculable rationality was seen as the opening for the other to come. To allow for a community open to the other to come, reason must be reasoned with, he said, while summoning a community whose energies derived from the incalculable but which only could open up to the other by those already capable of venturing into a reasoned engagement with its grounding rationale.

Thus, there are particularly some problems that stand out in the attempt to engage a community not predicated on a notion of sameness that presupposes difference. First is the problem of a linear understanding of time and a homogeneous view of space that serve as frameworks for imagining the community. Second is the problem of a *logos*, or an emphasis on political speech and visibility, as the measure to grasp beings as political and as part of the community. Third is the problem of making representational thinking the ground for the delineation of the community. Fourth is the problem of making boundaries the condition of possibility for politics, since this makes a community predicated on a notion of sameness premised on difference necessary for the continuation of politics. Finally is the problem of a democratic mastery that counts, since any count requires a measure for who is to be counted.

What relates these problems is that they all hinge on representational or calculative thinking. Not even the problem of politics as contestation escaped this, since contestations with sources in incalculability eventually were consumed in new measures for calculation. Now, what relates representation and calculation is that *they both rely on mastery*. Thereby we are back at the question of the symptom of statelessness as a symptom.

This book has suggested that Arendt's observation of the stateless as the most symptomatic group in contemporary politics could be read with respect to a community predicated on sameness and difference. Doing so, the stateless were said to form a symptom of the boundaries at work in the community. They were seen as a symptom of how communal belonging was established in terms of a membership that overlooked the political existence of those lacking such membership. Finally, the stateless were seen as symptomatic of a closure that narrowed down the very meaning of politics. Thereby, the predicament of the stateless transpired also to the ones formally included in the community.

The book argued that seeing statelessness as a symptom formed a symptom in itself. It formed a symptom of a rationale that gave priority to the state and to a certain kind of individual seen as fitting to the state. Such a rationale, it was said, was found in calculability. This, we have now seen, is intrinsically linked to a community predicated on sameness. It forms not only part of a nationalist thinking that seeks to ground membership on bloodlines. Calculability forms also a paradoxical part of democracy and runs always the risk of undermining the very meaning of a democratic people.

What happens, then, with the question of political community in the encounter between the elaborations on *place as a difference without identity* and the *predicament of situation* as a sameness that does not presuppose difference, and the issues of democracy, mastery, and calculability? Chapter 5 suggested that the predicament of situation could be grasped as the problem of dwelling in the face of an authoritative count. The chapter also maintained that the predicament of situation did not answer the question of the political community; rather, it formed a premise for the normative orientation of the community, since the existential predicament and the political problem of an exclusive community predicated on sameness share the same source—calculability. However, in the introduction, it was said that statelessness as a symptom formed a symptom in itself—a symptom of a rationale that gives priority to the state and shows concern over the stateless rather than elaborating the meaning of the *statefree*. What remains of this book will engage the two realms unfolding from the perspectives taken on the problem of statelessness as either symptomatic of a community predicated on sameness or as a symptom in itself of a rationale the gives priority to the state. It is suggested that the predicament of situation and place as a difference without identity unfold along these two different registers of the question of the political community. The predicament of situation is formulated in response to statelessness as a symptom of contemporary politics and to the problem of grasping a democratic people in the face of democratic mastery. Thus, it describes the experience of the incalculable vis-à-vis an authoritative count. Place as a difference without identity, on the other hand, approaches the question of the political community on the register of the statefree. That is, it attempts to undo a rationale giving priority to the state and elaborates the meaning of an incalculable community and the way it reformulates and relocates the question of the political community.

The remaining two parts engage the predicament of situation and place as difference without identity respectively, in order to grasp how these two figures restate the issues of membership, territory, and politics.

II

It was said that being in place is not an easy task. On the contrary, to dwell emerged as something we continuously must learn again. I suggested this gave rise to a notion of sameness that does not presuppose difference. This was said to be found in a predicament of situation that describes the shared problem of being torn between *situation*, that is, the lived, manifold experience of being in place, and a *position* in which a calculable reason orders beings according to categorisation. Sameness as a predicament of situation does not give rise to a unified us. Although some of those sharing in the predicament of situation might be part of communities predicated on a notion of sameness presupposing difference, this is not the case for all. To the vaster part of those who share in this predicament, the sameness of positioning is only felt, not as a resource for stability, security, or participation, but as encroaching on the possibilities for existence. Further, the sameness of sharing in the predicament of situation does *not* exactly describe the place of the community. What is shared in the predicament of situation might at large be to merely be positioned, and thus to engage with other beings through sameness and identification, which is the opposite to the encounters of difference inherent to an incalculable community. Still, since the predicament of situation is something that we all experience as far as we are affected by the reach of modern space and calculation, it forms a sameness that does not give rise to difference. What, then, are the normative and political implications of this? And especially, what are the implications for belonging, space, and the meaning of politics?

The last chapter saw how the calculable/incalculable copula brought moral commitments into the realm of *logos*. Consequently, a revalorisation of humanity was seen to emerge not only for Heidegger but also for Derrida. Although the predicament of situation arguably would run the same risk, it was argued that if it is through place that beings are part of world building and of forming an incalculable community, then the only moral precept is to let beings be in place. Such a moral principle cannot be fostered through, for example, a technical measure of equality developed in the realm of *logos*, since this would be external to place and encumbering on its difference. Such a moral principle would rest on a moral *concern* that the self carries out vis-à-vis the other. The notion of place carved out in this book suggests instead a moral *sparing* of the other that lets the other be in its own unfolding and gathering of space. This implies a wondrous attentiveness to how beings emerge in place and how oneself as well as the community unfolds with them. Since space is always joined and reciprocal, sparing the other from an invasive sameness and positioning means to also spare oneself from positions, determinations, and categorisations. Thus, sparing means to reduce the hold of sameness and identification in any encounter, not merely with the other but also with oneself. As a normative community, the incalculable community consists in the mutual relation of sparing, through which beings reciprocate in letting each other be. Doing so, they reject the sameness that stands in the way for coexistence—a notion of sameness that can be found in belonging understood as membership—and are thus engaged in an encounter formative both of themselves and their world.

Further, this reciprocity is not geographically limited. By contrast, the river draws the place of the community as never containable in a specific area or within a linear unfolding of time. Any separation of the regional from the global is thereby disrupted. Moral commitments that take 'river place' as their point of departure involve attending to how any formations of what is taken to be here have effects that reach also elsewhere, since river place undoes the possibility to determine a here without it also being related to places elsewhere. What occurs elsewhere is inescapably entwined with what happens here. One can thus not simply engage in a delineated community separated from other communities or from a larger whole of equally determined communities. Instead, the place of the community developed in this book shows how—no matter if geographically distant or close—places are adjoined and bleed into each other. So do moral commitments, which can never only be directed towards what is geographically close. Instead, moral commitments need to disrupt the relation between here and there that secures these as apart and instead interrogate the hold sameness and calculability have both here and elsewhere. The questions theories on citizenship have struggled with—where and how to draw the boundaries of moral concern—are thereby transformed into a question of how to engage in a moral sparing that reaches beyond the immediately close. The predicament of situation has showed how the community is not consolidated by the assimilation or integration of newcomers under either a common identity or a particular category. Instead, integration is transformed from being a question of how to incorporate newcomers into an existing community into a task that every being of that community has to engage with. The possibility of coexistence rests in a continued exchange without appropriation between beings, and the predicament of situation reveals how we are all in the same boat in having to engage in sparing to be able to set ourselves free to the encounters that bring the community forth.

Democratic theory revealed the problem of an incalculable *demos*, on the one hand, and, on the other, a demo*cracy* that requires a count. Hence, despite attempts to move away from state-centred notions of community and from the lack of democratic legitimacy of boundaries, democratic theory was seen to hinge on a state rationality of calculability. *Since democratic theorists in some way or another confess to equality, they are bound to give way to distinct communities.* Without necessarily having the nation-state as the ground for democracy, they still invite a state logic, be its claims local or global, as a foundation for democracy. Now, the problem with this, the book suggested, is that equality becomes conflated with sameness. Whether equality in number or worth, equality rests on an assumption of sameness. Equality of number counts each subject once but has to ascertain that a subject indeed is countable as a subject. Equality as worth seeks a measure for what counts as *logos*. Both notions of equality, however, invite sameness through the back door, leaving demo*cracy* to eventually undo the possibility of an incalculable *demos*.

Thus, democratic theory relies on and lets politics be structured around boundary setting by establishing equality as inclusion while giving way to distinctions between us and them. However, sameness as a predicament of situation disrupts this by showing that *there is no boundary from which such a distinction can be made*. No matter if safely within the confines of a particular state or if dwelling on its margins, as long as we are affected by the reach of modernity and calculability, we all are exposed to and have to engage a predicament of situation. Therefore, the task for future studies of democratic theory appears to be to either develop a measure for the incalculable, some of the difficulties of which have already been touched upon, or to seek to grasp a notion of equality that does not presuppose sameness.

To be noted, this would not give way for a neoliberal or even anarcho-capitalist rationale. To the contrary, a neoliberal notion of ownership would align with the understanding of space as measurable, calculable, and bounded that this book has sought to eschew. If place is for all, then space cannot be owned. This has far-reaching consequences for the individual as well as for the state, since it contests the exclusivity of land. A democracy seeking to live up to its aspirations—the rule of the manifold—needs therefore not only to be spatially unbounded but also function to unbind space from its calculable connotations.

Rather, a certain equality that does not form a measure could be discerned in the notion of place as a difference without identity. Let us follow this path when taking up the question of mastery. The book has confirmed Elden's observation that "there is a politics of space because *politics is spatial*" by showing how space is implicated in the problem of representational thinking and of presuming sameness as a condition of possibility for the community. Projected as smooth, bounded, and owned or inherited, space has furthered a unified image of the people. Thus, a politics directed to the ordering and unification of space has been put to practice; it has bounded space, fragmentised it, and created hierarchies within it, all under the guiding light of sameness.

If politics is spatial and if, as seen in this book, sovereignty presumes a politics of space that orders according to sameness and positions, then taking the predicament of situation seriously must have severe implications for sovereignty. From the point of view of the sameness of space, there is little difference between the sameness of nation-state sovereignty, which Derrida saw as a possibility to interrupt global claims to power, and the sameness of, for example, global liberalism. Both the nation-state and liberal universalism posit a sameness of space, claim a capacity to survey it, and mark out the positions possible within that space. Thus, they both rely on space as calculable and bounded in reach or in meaning.

Derrida's resolution to the problem of a calculating sovereign reason was to reason with reason, so as to make way for another, incalculable rationale. The task for reason was to open up to the incalculable and fully other and for a democracy to come. I, however, maintained that while the transaction between calculable and incalculable reason would augment the meaning of community, it also inescapably remained within reason. Consequently, the other emerged only with the concern showed by the ones already included in that community. To include the other, a reasoned transaction between the calculable and the incalculable needed to be carried out, though since this transaction kept within reason, and since the other was fully other to the *logos* of the community, it followed that the route to open up towards the incalculable went through a calculable inside. The other remained as stateless, as a being that failed being grasped as political in her own right.

To instead see the political as spatial, and to grasp place as a notion of difference that renounces an underlying identity, will recast this problem. Place as difference without identity reveals that to make room for the incalculable, if the incalculable at all should be able to enter into Derrida's "autoimmune aporia" of the impossible transaction between the calculable and the incalculable, then *sovereignty must already be unconditionally conditioned on the possibility of an incalculable place*. I have here argued that incalculability enters the world through place, not through language, not through political disagreement, not through enacting a position or claiming a certain rationale. First of all, it emerges in and with the world through place. Any aspirations to a democratic sovereignty thus abandon claims to territorial integrity, as well as to the ability to project a homogeneous and transparent space. Thus, to the extent democratic mastery can persist, claims to territorial ownership must be relinquished, as must both the external and internal borders intending to separate citizens from those kept out of place. Instead, *sovereignty must be conditioned on leaving room for place and on relating to beings through "letting dwell"*.

Sameness as a predicament of situation therefore restates the question of political community. It is no longer oriented to the issue of who and according to what characteristics. Rather, it forms a question of how to maintain the openness to what is strange and incalculable. Honig noticed how democracy is to live with strangers under a law that is therefore alien. However, this conclusion is dependent on the idea of the democratic community as delineated *prior* to the emergence of what is strange to it. To instead engage the equality of a notion of place that describes a difference without identity means to also engage boundaries differently, not as where the community ends, but as where it begins its presencing. Oriented to the incalculability of place, belonging comes with any form of dwelling in difference without identity, rather than as a membership conditioned on a prior ascription to a notion of sameness predicated on difference. Sameness as a predicament of situation also restates the spatiality of the community. It cannot be grasped as the smooth, homogeneous space of the territory, since this would not leave room for the incalculability of place. Instead, it opens a volatile, ephemeral spatiality that encroaches on the overlap between sovereignty and space and disqualifies any territorial ownership or exclusivity. The political implications of this are immense, as they intervene in the very foundations of sovereignty.

They also intervene in our understanding of democratic mastery as relying on the authority of a count. However, does the count of democracy form a rationale necessary to democracy or is it a rationale necessary to the state? Remaining with the state, we are bound to remain with a politics structured around boundaries, as, it appears, the democratic *state* must delimit the plurality and incalculability of the democratic people. Thus, to grasp the *meaning of the political*, we need to venture further into the statefree and the meaning of an incalculable community.

III

Place as a notion of difference without identity—both in terms of the place of the community and of the singular being—undoes many of the problems of a community that takes an understanding of sameness predicated on difference as its condition of possibility. River place breaches the problem of linear time, both in the sense of a linear teleology of progress and in the sense of a nostalgic projection of a lost origin. By its sheer difference without identity, place will always vest the community with endless possibilities of what is to come. In addition, the river showed how previous locales have already been altered by the locales that follow. Now, this could be grasped as if what is here has altered places past and transformed them through a, perhaps, nostalgic projection, as would be the case of a nationalist thinking that seeks to recover the places presumably once had in all their imagined glory. However, such thinking projects a sameness and identity to a place passed, indeed, a place that never was. Nostalgic projection would thus go against the entire meaning of place as difference without identity. Still, when here and there connects to a now and then, one must be on guard against the ghost of nostalgia as well as of teleology and instead bear in mind how the path of the river itself unfaithfully abandons the human landscape. Instead, the time of the river will require bringing any places passed into a new light by intimating them anew in yet unknown configurations.

Place thereby alters the meaning of boundaries. Marked by the river, and formed by beings who are in place through emerging in and with the communal place, an incalculable community does not easily give way to boundaries. To the extent that it is at all possible to talk of any boundaries of the community, these are not located in space but rather in a close past. If a boundary, a *horismos*, is there from which something begins its presencing, and if each being emerges together with a place and in so doing is already there, then the boundaries of the community are always already behind us. When a being presences in the community, this does not form a boundary that the community needs to orient itself towards, engage, and potentially overcome so as to count the being in question as included. On the contrary, each interaction in the community has already altered its place and has thus already participated in the presencing of the community. Hence, the community orientates itself *after* that boundary and the new horizons for common existence unfolding from it. Forming horizons, boundaries not only let spaces unfold but retreat also in the distance. If there are boundaries to the community, these are in a non-locatable past; they have already vanished with the flow of the river. When transgressing boundaries becomes something that cannot be undone, it could be suspected that some would call for a fortification of borders to make sure that everything unwanted is kept out. But building walls implies sticking with a view of space as a boundable, as a container, and as sameness, which would preclude the possibility of community altogether.

Dwelling place interrupts the centrality of speech, logos, and visibility for perceiving beings as part of the community. By simply being there, beings are part in gathering the place of the community. Whether in public or out of sight in the nooks and crannies of the hidden corners of the world, they already take part in the building of a world and the emergence of the community. The removal of a being from the place of the community does not simply form an erasure of any trace of her from the map of the community. The possibility of such a removal relies on a logic of sameness that assumes space as a mere container of people. From the perspective of dwelling, the removal of a being will, on the contrary, cut a wound in the community, cripple it, and bring its flowing place to a halt. The fate of a being is interwoven in the fabric of the community. Whether still physically present or not, they form part of its historical unfolding. Imprisoned asylum seekers, evicted irregular migrants, refugees drowning on their way to an uncertain and temporary security, or unaccompanied refugee children living a dangerous clandestine existence to escape expulsion are not aberrations or externalities to an otherwise smoothly operating society. They are not a temporary crisis. On the contrary, exposing the community to the injustices, atrocities, and violence in its wake, they are crucial to the continued turn of the *polis* and the presencing of a common place. Whatever their formal status, as dwelling, they already form part of the community and its place and history. Even when made absent, dwelling beings have already entered into and altered the community's appearance. They have affected how the community comes to presence and inscribed brutality, repression, and violence as undeniable parts of the community. The notion of river space forces the community to return to its past and once more bring it to light and grasp how predicaments of today affect and are affected by events in the past. Thus, with river space, the world unfolds as an integrated and interwoven place where historical unfolding forms a dense web of relations of interdependence.

Hence, whereas the emphasis on speech and *logos* retained an inequality at the heart also of the democratic community, there is an equality to place since it rejects relations of dependence, visibility, and speech as necessary for world building. Instead, place is incalculable and for all. A being is a being in place; before any language, before articulating claims of her common existence, she has—as a being in place—already acted. Therefore, the incalculable community stemming from place as a difference without identity undoes the problem of statelessness as symptomatic to how being part in the community takes the form of a membership. *Where there is place for all, no membership is needed.*

Determining the place of the human abode, the river does indeed the opposite to *determine* the place of the community when revealing representations as a secondary and always reductive undertaking. Instead, the river vests the place of the community with a flow and ephemerality *that evades any representational thinking*. As a difference that does not presuppose identity, place draws a notion of 'us' that does not first emerge through running aground in a particular meaning. By contrast, it suggests a community before any recourse to language. If the locality of the river forms an interplay of places, then it suggests that places touch each other, are joined, and already form part of that abode which now might no longer be marked out as merely human. An incalculable community interrupts and disqualifies the understanding of space as territory. Instead of a space vested in sameness, the incalculable community finds its very source in the situated experience of dwelling or being in place, an experience marked by the inability to be appropriated according to a particular rationale.

Thus, an incalculable community of statefree beings does not unfold from any above imposed principles for ordering, nor does it exist delimited by any particular geographical boundary. Instead *the incalculable community relocates the community to unfold between beings*. As dwelling, statefree beings are already part in the incessant gathering of the place of the community. Immersed in the landscape, they share in the presencing of a place and community that is not bounded. They encounter each other through an openness that follows from letting go of identification and sameness as grounds for engagement with the other. Instead, they share in their inability to be appropriated, neither by the other nor by themselves. As dwelling, statefree beings are already part of an incalculable community where there is place for all.

What are, then, the implications of the incalculable community for the meaning of the political? This book has seen politics emerge in two different forms: one in which politics orders a particular community, group, or worldview; the other, agonist understanding, saw how politics forms a contestation of such a way of ordering. Both these forms rely on a notion of sameness that presupposes difference, the first through assuming sameness as the condition of possibility of the community, the second through assuming sameness as the condition of possibility of political contestation. Thus, to ask what a politics would mean that does not assume sameness as its condition of possibility is no easy question. A hint might be given by the river, which saw the place of the community emerge as transitory, multifarious, and volatile. If, as Heidegger says, the political must be understood on the basis of the meaning of *polis*, then we need to search for a meaning of a politics that takes transience, ephemerality, and difference without identity as its condition of possibility. This means to *attend to the politics emanating from the shared inability to be appropriated*, an inability shared between human and non-human beings alike, as well as between beings and their world.

Elden's reformulation of Lefebvre's observation of how "there is a politics of space because space is political" into the insight that "there is a politics of space because politics is spatial" means also a relocation of the ontological register on which space is analysed. While space certainly, for Lefebvre, is ontologically felt, the statement installs space on the level of the ontic. Space is what politics makes of it; it becomes ordered according to the rationale that orders the meaning of politics and grasped only according to this rationale. According to the first statement, space therefore becomes instilled with boundaries, ownership, fragmentation, and hierarchies. The aim of politics is to order places, in accordance with its own logic, including the conceptualisation, delimitations, and boundary setting this rests on.

In the other case, that sees a politics of space because politics is spatial, the spatial unfolds on the level of the ontological, of what makes possible a situated existence. This does not exclude the possibility of an ontic politics of space that seeks to control, contain, and eventually close off the incalculability of place. However, it also draws another meaning of the political that sees any appearance of place stemming from any existential situation—be it the fox's lair, the vulnerability of a clandestine life, or the public staging of an assumed equality—form part to the continued questioning of the community. If the political assumes the form of a situation, as suggested in the previous chapter, then any lived experience of being in place is instilled with and forms an expression of political agency. Any lived existence reaches into the spatiality of the community and alters the meaning and possibilities of the community. By being there, it forms already a situated part in the historical and spatial unfolding of the political community.

Approaching the question of political community through the incalculability of place means to engage the spatiality of the question itself, or rather, *questioning as situation*. As a question, place as a difference without identity draws a belonging not predicated on membership or even the goodwill of others. As a question, place intervenes in the spatiotemporal boundedness of the political community and expands its meaning and possibilities. As situation, place draws a political community already there, as already present in its presencing of a site gathered anew. Approaching the question of political community through the incalculability of place implies therefore the *relocation* of the question of the political community that locates the community as unfolding with each being dwelling in place. It means also a *reformulation* of the question of the political community, through attending to the political community, political agency, and belonging as situation, rather than as posited, bounded existence. Here, the meaning of a statefree existence begins to take form. When approached from the point of view of sovereignty, of the state, or from a democratic theory that despite its ambitions seems to conjure so many remainders, this might seem like a distant dream of a community perhaps to come. When approached through attending to those already part in the presencing of the place of the community, the question of the incalculable community rather suggests the openness to a community already *there*.

References

Abizadeh, Arash. 2008. "Democratic Theory and Border Coercion: No Right to Unilaterally Control Your Own Borders." *Political Theory* 36 (1): 37–65.

———. 2012. "On the Demos and Its Kin: Nationalism, Democracy, and the Boundary Problem." *American Political Science Review* 106 (4): 867–82.

Agnew, John. 1994. "The Territorial Trap: The Geographical Assumptions of International Relations Theory." *Review of International Political Economy* 1 (1): 53–80.

Anderson, Ben. 2008. "For Space (2005): Doreen Massey." In *Key Texts in Human Geography: A Reader Guide*, edited by Phil Hubbard, Rob Kitchin, and Gill Valentine, 227–35. London: Sage.

Anderson, Benedict. 2006. *Imagined Communities: Reflections on the Origin and Spread of Nationalism*. London: Verso.

Archibugi, Daniele, and David Held. 1995. *Cosmopolitan Democracy: An Agenda for a New World Order*. Cambridge: Polity Press.

Arendt, Hannah. 1976. *The Origins of Totalitarianism*. San Diego: Harcourt.

———. 1998. *The Human Condition*. Chicago: Chicago University Press.

Bachelard, Gaston. 1994. *The Poetics of Space*. Boston: Beacon Press.

Badiou, Alain. 2005. *Metapolitics*. London: Verso.

Balibar, Étienne. 2002. *Politics and the Other Scene*. London: Verso.

Bartelson, Jens. 1998. "Second Natures: Is the State Identical with Itself?" *European Journal of International Relations* 4 (3): 295–326.

———. 2001. *The Critique of the State*. Cambridge: Cambridge University Press.

———. 2009a. "Globalizing the Democratic Community." In *The Future of Political Community*, edited by Gideon Baker and Jens Bartelson, 36–50. Milton Park, UK: Routledge.

———. 2009b. *Visions of World Community*. Cambridge: Cambridge University Press.

———. 2014. *Sovereignty as Symbolic Form*. London: Routledge.

Behnke, Andreas. 1997. "Citizenship, Nationhood and the Production of Political Space." *Citizenship Studies* 1 (2): 243–65.

Beistegui, Miguel de. 1998. *Heidegger and the Political*. New York: Routledge.

———. 2003. *Thinking with Heidegger: Displacements*. Bloomington: Indiana University Press.

———. 2011. "The Place of Place in Heidegger's Topology." *International Journal of Philosophical Studies* 19 (2): 277–83.

Benhabib, Seyla. 2004. *The Rights of Others: Aliens, Residents, and Citizens*. Cambridge: Cambridge University Press.

———. 2006. "Another Cosmopolitanism." In *Another Cosmopolitanism: The Berkeley Tanner Lectures*, edited by Robert Post. New York: Oxford University Press.

Benton, Lauren. 2010. *A Search for Sovereignty: Law and Geography in European Empires, 1400–1900*. Cambridge: Cambridge University Press.

Bhabha, Homi K. (ed.). 1990. *Nation and Narration*. London: Routledge.

———. 2004. *The Location of Culture*. London: Routledge.

Bingham, Charles W., and Gert Biesta. 2010. *Jacques Rancière: Education, Truth, Emancipation*. London: Continuum.

Birnbaum, Daniel, and Sven-Olov Wallenstein. 1999. *Heideggers väg*. Stockholm: Thales.

Bohman, James. 2010. *Democracy across Borders: From Dêmos to Dêmoi*. Cambridge, MA: MIT Press.

Branch, Jordan. 2011. "Mapping the Sovereign State: Technology, Authority, and Systemic Change." *International Organization* 65 (1): 1–36.

Brenner, Neil, and Stuart Elden. 2009. "Henri Lefebvre on State, Space, Territory." *International Political Sociology* 3 (4): 353–77.

Brown, Wendy. 2010. *Walled States, Waning Sovereignty*. New York: Zone.

———. 2015. *Undoing the Demos: Neoliberalism's Stealth Revolution*. Cambridge, MA: MIT Press.

Brubaker, Rogers. 1992. *Citizenship and Nationhood in France and Germany*. Cambridge, MA: Harvard University Press.

Canovan, Margaret. 2005. *The People*. Cambridge: Polity Press.

Casey, Edward S. 1993. *Getting Back Into Place: Toward a Renewed Understanding of the Place-World*. Bloomington: Indiana University Press.

———. 1997. *The Fate of Place: A Philosophical History*. Berkeley: University of California Press.

Cavallar, Osvaldo. 2004. "River of Law: Bartolus's 'Tiberiadis' ('De alluvione')." In *A Renaissance of Conflicts: Visions and Revisions of Law and Society in Italy and Spain*, edited by John A. Marino and Thomas Kuehn. Toronto: Centre for Renaissance and Reformation Studies.

Cloke P., and O. Jones. 2001. "Dwelling, Place, and Landscape: An Orchard in Somerset." *Environment and Planning A* 33 (4): 649–66.

Connolly, William E. 1991. *Identity/Difference: Democratic Negotiations of Political Paradox*. Minneapolis: University of Minnesota Press.

———. 1995. *The Ethos of Pluralization*. Minneapolis: University of Minnesota Press.

———. 2005. *Pluralism*. Durham, NC: Duke University Press.

Corcoran, Steve. 2010. "Foreword." In *Dissensus: On Politics and Aesthetics*, written by Jacques Rancière. London: Continuum.

Couture, Jean-Pierre. 2009. "Spacing Emancipation? Or How Spherology Can Be Seen as a Therapy for Modernity." *Environment and Planning D: Society and Space* 27 (1): 157–63.

Coward, Martin. 2008. *Urbicide: The Politics of Urban Destruction*. New York: Routledge.

Craven, Matthew. 2006. "The Violence of Dispossession: Extra-Territoriality and Economic, Social, and Cultural Rights." In *Economic, Social, and Cultural Rights in Action*, edited by Mashood Baderin and Robert McCorquodale, 71–88. Oxford: Oxford University Press.

Deleuze, Gilles. 2004. *Difference and Repetition*. London: Continuum.

Derrida, Jacques. 1982. *Margins of Philosophy*. Chicago: University of Chicago Press

———. 1989. *Of Spirit: Heidegger and the Question*. Chicago: University of Chicago Press.

———. 1993. "Heidegger's Ear: Philopolemology." In *Reading Heidegger: Commemorations*, edited by John Sallis. Bloomington: Indiana University Press.

———. 1996. *Monolingualism of the Other; or, The Prosthesis of Origin*. Stanford, CA: Standford University Press.

———. 2001. *On Cosmopolitanism and Forgiveness*. London: Routledge.

———. 2005. *Rogues: Two Essays on Reason*. Stanford, CA: Stanford University Press.

Douzinas, Costas. 2007. *Human Rights and Empire: The Political Philosophy of Cosmopolitanism*. Abingdon, UK: Routledge.

Duffield, Mark. 2006. "Racism, Migration and Development: The Foundations of Planetary Order." *Progress in Development Studies* 6 (1): 68–79.

Edensor, Tim. 2006. "Reconsidering National Temporalities: Institutional Times, Everyday Routines, Serial Spaces and Synchronicities." *European Journal of Social Theory* 9 (4): 525–45.

Elden, Stuart. 1999. "Heidegger's Hölderlin and the Importance of Place." *Journal of the British Society for Phenomenology* 30 (3): 258–74.

———. 2000. "Rethinking the Polis: Implications of Heidegger's Questioning the Political." *Political Geography* 19 (4): 407–22.

———. 2001. *Mapping the Present: Heidegger, Foucault and the Project of a Spatial History*. London: Continuum.

———. 2004. *Understanding Henri Lefebvre: Theory and the Possible*. London: Continuum.

———. 2006. *Speaking against Number: Heidegger, Language and the Politics of Calculation*. Edinburgh: Edinburgh University Press.

———. 2009. *Terror and Territory: The Spatial Extent of Sovereignty*. Minneapolis: University of Minnesota Press.

———. 2010a. "Thinking Territory Historically." *Geopolitics* 15: 757–61.

———. 2010b. "Thinking Territory Politically." *Political Geography* 29 (4): 238–41.

———. 2013a. *The Birth of Territory*. Chicago: University of Chicago Press.

———. 2013b. "How Should We Do the History of Territory?" *Territory, Politics, Governance* 1 (1): 5–20.

———. 2014. "Interview with Gastón Gordillo, author of *Landscapes of Devils* and the forthcoming *Rubble: The Afterlife of Destruction*." societyandspace.com/material/interviews/interview-with-gaston-gordillo-author-of-landscapes-of-devils-and-the-forthcoming-rubble-the-afterlife-of-destruction/.

Erskine, Toni. 2000. "Embedded Cosmopolitanism and the Case of War: Restraint, Discrimination and Overlapping Communities." *Global Society: Journal of Interdisciplinary International Relations* 14 (4): 569–90.

———. 2002."'Citizen of Nowhere' or 'The Point Where Circles Intersect'? Impartialist and Embedded Cosmopolitanisms." *Review of International Studies* 28 (3): 457–78.

Espejo, Paulina Ochoa. 2011. *The Time of Popular Sovereignty: Process and the Democratic State*. University Park: Pennsylvania State University Press.

———. 2014. "People, Territory, and Legitimacy in Democratic States." *American Journal of Political Science* 58 (2): 466–78.

Esposito, Roberto. 2010. *Communitas: The Origin and Destiny of Community*. Stanford, CA: Stanford University Press.

Fasolt, Constantin. 2004. *The Limits of History*. Chicago: University of Chicago Press.

Fielding, Helen A. 2008. "Dwelling with Language: Irigaray Responds." In *French Interpretations of Heidegger: An Exceptional Reception*, edited by David Pettigrew and Francois Raffoul. Albany: State University of New York Press.

Fitzpatrick, Peter. (2004). Terminal Legality? Human Rights and Critical Being. In *Critical Beings: Law, Nation and the Global Subject*, edited by Fitzpatrick, Peter and Tuitt, Patricia, 119–136. Hants: Ashgate.

Foucault, Michel. 1982. "The Subject and Power." *Critical Inquiry* 8 (4): 777–95.

Gregory, Derek. 2004. *The Colonial Present: Afghanistan, Palestine, and Iraq*. Malden, MA: Blackwell.

Grotius, Hugo. 2000. *The Rights of War and Peace: Including the Law of Nature and of Nations*. Chestnut Hill, MA: Elibron.

Gunneflo, Markus, and Niklas Selberg. 2010. "Discourse or Merely Noise? Regarding the Disagreement on Undocumented Migrants." *European Journal of Migration and Law* 12 (2): 173–91.

Harvey, David. 1996. *Justice, Nature and the Geography of Difference*. Oxford: Blackwell.

Heater, Derek. 1996. *World Citizenship and Government: Cosmopolitan Ideas in the History of Western Political Thought*. Basingstoke, UK: Macmillan.

———. 2002. *World Citizenship: Cosmopolitan Thinking and Its Opponents*. London: Continuum.

Heidegger, Martin. 1969. *Identity and Difference*. Chicago: University of Chicago Press.

———. 1973. "Art and Space." *Man and World* 6: 3–8.

———. 1992. *Parmenides*. Bloomington: Indiana University Press.

———. 1996a. *Bemerkungen zu Kunst-Plastik-Raum*. St. Gallen: Erker Verlag.

———. 1996b. *Hölderlin's Hymn "The Ister."* Bloomington: Indiana University Press.

———. 2001a. "Building Dwelling Thinking." In *Poetry, Language, Thought*, written by Martin Heidegger. New York: Perennical Classics.

———. 2001b. "The Origin of the Work of Art." In *Poetry, Language, Thought*, written by Martin Heidegger. New York: Perennical Classics.

———. 2002a. "The Age of the World Picture." In *Off the Beaten Track*, edited by Julian Young and Kenneth Haynes. Cambridge: Cambridge University Press.

———. 2002b. "Anaximander's Saying." In *Off the Beaten Track*, edited by Julian Young and Kenneth Haynes. Cambridge: Cambridge University Press.

———. 2010. *Being and Time*. Albany: State University of New York Press.

Held, David. 2002. "Culture and Political Community: National, Global, and Cosmopolitan." In *Conceiving Cosmopolitanism: Theory, Context and Practice*, edited by Steven Vertovec and Robin Cohen, 48–58. Oxford: Oxford University Press.

Hesse, Hermann. 1951. *Siddhartha*. New York: New Directions.

Hindess, Barry. 2000. "Citizenship in the International Management of Populations." *American Behavioral Scientist* 43 (9): 1486–97.

Honig, Bonnie. 2001. *Democracy and the Foreigner*. Princeton, NJ: Princeton University Press.

——— 2006. "Another Cosmopolitanism? Law and Politics in the New Europe." In *Another Cosmopolitanism: The Berkeley Tanner Lectures*, edited by Robert Post. New York: Oxford University Press.

———. 2009. *Emergency Politics: Paradox, Law, Democracy*. Princeton, NJ: Princeton University Press.

———. 2013. *Antigone, Interrupted*. Cambridge: Cambridge University Press.

Ingold, Tim. 1993. "The Temporality of the Landscape." *World Archaeology* 25 (2): 152–74.

———. 1995. "Building, Dwelling, Living: How Animals and People Make Themselves at Home in the World." In *Shifting Contexts: Transformations in Anthropological Knowledge*, edited by Marilyn Strathern. London: Routledge.

———. 2005. "Epilogue: Towards a Politics of Dwelling." *Conservation and Society* 3 (2): 501–8.

Irigaray, Luce. 1993. *An Ethics of Sexual Difference*. Ithaca, NY: Cornell University Press.

———. 1999. *The Forgetting of Air in Martin Heidegger*. Austin: University of Texas Press.

Isin, Engin F. 2002. *Being Political: Genealogies of Citizenship*. Minneapolis: University of Minnesota Press.

———. 2007. "City.State: Critique of Scalar Thought." *Citizenship Studies* 11 (2): 211–28.

———. 2009. "Citizenship in Flux: The Figure of the Activist Citizen." *Subjectivity* 29 (1): 367–88.

James, Ian. 2012. *The New French Philosophy*. Cambridge: Polity.

Janicaud, Dominique. 1993. "The 'Overcoming' of Metaphysics in the Hölderlin Lectures." In *Reading Heidegger: Commemorations*, edited by John Sallis. Bloomington: Indiana University Press.

Jones, Reece. 2009. "Categories, Borders and Boundaries." *Progress in Human Geography* 33 (2): 174–89.

Kolb, David. 1990. Postmodern *Sophistications: Philosophy, Architecture and Tradition*. Chicago: University of Chicago Press.

Kostakopoulou, Theodora. 2009. "Citizenship Goes Public: The Institutional Design of Anational Citizenship." *Journal of Political Philosophy* 17 (3): 275–306.

Krell, David Farrell. 1993. "Spiriting Heidegger." In *Of Derrida, Heidegger, and Spirit*, edited by David Wood. Evanston, IL: Northwestern University Press.

Kymlicka, Will. 1995. *Multicultural Citizenship: A Liberal Theory of Minority Rights*. Oxford: Clarendon Press.

———. 2001. *Politics in the Vernacular: Nationalism, Multiculturalism and Citizenship*. Oxford: Oxford University Press.

Lacoue-Labarthe, Philippe, and Jean-Luc Nancy. 1997. *Retreating the Political*. London: Routledge.

Land, Nick. 1993. "Spirit and Teeth." In *Of Derrida, Heidegger, and Spirit*, edited by David Wood. Evanston, IL: Northwestern University Press.

Lefebvre, Henri. 1991. *The Production of Space*. Oxford: Basil Blackwell.

———. 2009. *State, Space, World: Selected Essays*. Minneapolis: University of Minnesota Press.

Lindahl, Hans. 2014. *Fault Lines of Globalization: Legal Order and the Politics of A-Legality*. Oxford: Oxford University Press.

Lloyd, Moya. 2005. *Beyond Identity Politics: Feminism, Power and Politics*. London: Sage.

Macnaghten, Phil, and John Urry. 1998. *Contested Natures*. London: Sage.

Malpas, Jeff. 2006. *Heidegger's Topology: Being, Place, World*. Cambridge, MA: MIT Press.

———. 2012. *Heidegger and the Thinking of Place: Explorations in the Topology of Being*. Cambridge, MA: MIT Press.

Marchart, Oliver. 2007. *Post-Foundational Political Thought: Political Difference in Nancy, Lefort, Badiou and Laclau*. Edinburgh: Edinburgh University Press.

Markell, Patchen. 2003. *Bound by Recognition*. Princeton, NJ: Princeton University Press.

Marx, Karl. 1978. "On the Jewish Question." In *The Marx-Engels Reader*, edited by Robert Tucker, 26–46. New York: Norton.

Massey, Doreen. 1999. "Imagining Globalization: Power-Geometries of Time-Space." In *Global Futures: Migration, Environment and Globalization*, edited by Avtar Brah, Mary J. Hickman, and Máirtín Mac an Ghaill. Basingstoke, UK: Macmillan

———. 2005. *For Space*. London: Sage.

May, Stephen, and Chritsine E. Sleeter. 2010. *Critical Multiculturalism: Theory and Praxis*. New York: Routledge.

May, Todd. 2010. *Contemporary Political Movements and the Thought of Jacques Rancière: Equality in Action*. Edinburgh: Edinburgh University Press.

McFarlane, C. 2011. "The City as Assemblage: Dwelling and Urban Space." *Environment and Planning D: Society and Space* 29 (4): 649–71.

Modood, Tariq. 2007. *Multiculturalism: A Civic Idea*. Cambridge: Polity.

Mohanty, Chandra Talpade. 1984. *Under Western Eyes: Feminist Scholarship and Colonial Discourses* 12/13 (3/1): 333–358. *Boundary 2* (print).
Nancy, Jean-Luc. 1991. *The Inoperative Community*. Minneapolis: University of Minnesota Press.
———. 1993. *The Experience of Freedom*. Stanford, CA: Stanford University Press.
———. 2000. *Being Singular Plural*. Stanford, CA: Stanford University Press.
———. 2010. "Ändlig och oändlig demokrati." In *Vad innebär det att vara demokrat?* Stockholm: Tankekraft.
Näsström, Sofia. 2003. "What Globalization Overshadows." *Political Theory* 31 (6): 808–34.
———. 2007. "The Legitimacy of the People." *Political Theory* 35 (5): 624–58.
———. 2010. "Democracy Counts: Problems of Equality in Transnational Democracy." In *Transnational Actors in Global Governance*, edited by Jonas Tallberg and Christer Jönsson. Palgrave Macmillan.
———. 2011. "The Challenge of the All-Affected Principle." *Political Studies* 59 (1): 116–34.
Noll, Gregor. 2010. "Why Human Rights Fail to Protect Undocumented Migrants." *European Journal of Migration and Law* 12 (2): 241–72.
Norberg-Schulz, Christian. 1985. *The Concept of Dwelling: On the Way to Figurative Architecture*. New York: Rizzoli.
Nyers, Peter. 2003. "Abject Cosmopolitanism: The Politics of Protection in the Anti-Deportation Movement." *Third World Quarterly* 24 (6): 1069–93.
Odysseos, Louiza. 2003. "On the Way to Global Ethics?: Cosmopolitanism, 'Ethical' Selfhood and Otherness." *European Journal of Political Theory* 2 (2): 183–207.
Parekh, Bhikhu. 2006. *Rethinking Multiculturalism: Cultural Diversity and Political Theory*. Hampshire, UK: Palgrave Macmillan.
Prozorov, Sergei. 2009. "Generic Universalism in World Politics: Beyond International Anarchy and the World State." *International Theory* 1 (2): 215–47.
———. 2014a. *Ontology and World Politics: Void Universalism I*. London: Routledge.
———. 2014b. *Theory of the Political Subject: Void Universalism II*. London: Routledge.
Raffoul, Francois. 2012a. "The Event of Space—Review: Heidegger Among the Sculptors." *Gatherings: The Heidegger Cricle Annua*l 2: 89–106.
———. 2012b. "Review of Heidegger and the Thinking of Place: Explorations in the Topology of Being." *Notre Dame Philosophical Reviews*. ndpr.nd.edu/news/heidegger-and-the-thinking-of-place-explorations-in-the-topology-of-being/.
Rancière, Jacques. 1991. *The Ignorant Schoolmaster: Five Lessons in Intellectual Emancipation*. Stanford, CA: Stanford University Press.
———. 1999. *Disagreement: Politics and Philosophy*. Minneapolis: University of Minnesota Press.
———. 2004. *The Politics of Aesthetics: The Distribution of the Sensible*. London: Continuum.
———. 2007. *On the Shores of Politics*. London: Verso.

———. 2009. "Should Democracy Come? Ethics and Poltitics in Derrida." In *Derrida and the Time of the Political*, edited by Pheng Cheah and Suzanne Guerlac, 274–90. Durham, NC: Duke University Press.
———. 2010. *Dissensus: On Politics and Aesthetics*. London: Continuum.
Relph, Edward. 1976. *Place and Placelessness*. London: Pion.
Ricoeur, Paul. 1994. *Oneself as Another*. Chigago: University of Chicago Press.
Rose, Mitch. 2012. "Dwelling as Marking and Claiming." *Environment and Planning D: Society and Space* 30 (5): 757–71.
Ruin, Hans. 2013. *Frihet, ändlighet, historicitet: essäer om Heideggers filosofi*. Stockholm: Ersatz.
Saharso, Sawitri. 2003. "Feminist Ethics, Autonomy and the Politics of Multiculturalism." *Feminist Theory* 4 (2): 199–215.
Sallis, John. 1993. "Flight of Spirit." In *Of Derrida, Heidegger, and Spirit*, edited by David Wood. Evanston, IL: Northwestern University Press.
Savage, Robert. 2008. *Hölderlin after the Catastrophe: Heidegger, Adorno, Brecht*. Rochester, NY: Camden House.
Schatzki, Theodore R. 2007. *Martin Heidegger: Theorist of Space*. Stuttgart: Franz Steiner Verlag.
Scott, James C. 2009. *The Art of Not Being Governed: An Anarchist History of Upland Southeast Asia*. New Haven, CT: Yale University Press.
Shapiro, Michael J. 2000. "National Times and Other Times: Re-thinking Citizenship." *Cultural Studies* 14 (1): 79–98.
Sheller, M., and J. Urry. 2006. "The New Mobilities Paradigm." *Environment and Planning A* 38 (2): 207–26.
Shindo, Reiko. 2012. "Rethinking Community: Translation Space as a Departure from Political Community." *International Political Sociology* 6 (2): 149–64.
Shotter, John. 1993. "Psychology and Citizenship: Identity and Belonging." In *Citizenship and Social Theory*, edited by Bryan S. Turner. London: Sage.
Sigvardsdotter, Erika. 2012. *Presenting the Absent: An Account of Undocumentness in Sweden*. Uppsala: Uppsala Universitet.
Skinner, Quentin. 2002. "A Third Concept of Liberty." *Proceedings of the British Academy* 117: 237–68.
Sloterdijk, Peter. (2011). *Spheres Volume I: Bubbles Microspherology*. Los Angeles: Semiotext(e).
Soja, Edward W. 1989. *Postmodern Geographies: The Reassertion of Space in Critical Social Theory*. London: Verso.
Spivak, Gayatri Chakravorty. 1990. *The Post-Colonial Critic: Interviews, Strategies, Dialogues*. New York: Routledge.
Springer, Simon. 2012. "Anarchism! What Geography Still Ought to Be." *Antipode* 44 (5): 1605–24.
Stambaugh, Joan. 1969. "Introduction." In *Identity and Difference*, written by Martin Heidegger. Chicago: University of Chicago Press.
Stephens, Angharad Closs. 2010. "Citizenship without Community: Time, Design and the City." *Citizenship Studies* 14 (1): 31–46.

———. 2013. *The Persistence of Nationalism: From Imagined Communities to Urban Encounters*. London: Routledge.

Stevens, Jacqueline. 2010. *States without Nations: Citizenship for Mortals*. New York: Columbia University Press.

Strauss, Leo. 1959. *What Is Political Philosophy? And Other Studies*. Chicago: University of Chicago Press.

Vallega, Alejandro A. 2003. *Heidegger and the Issue of Space: Thinking on Exilic Grounds*. University Park: Pennsylvania State University Press.

Vattel, Emer de. 2008. *The Law of Nations: Or, Principles of the Law of Nature, Applied to the Conduct and Affairs of Nations and Sovereigns, With Three Early Essays on the Origin and Nature of Natural Law and on Luxury*. Indianapolis: Liberty Fund.

Vaughan-Williams, Nick. 2009. *Border Politics: The Limits of Sovereign Power*. Edinburgh: Edinburgh University Press.

Walker, R. B. J. 1993. *Inside/Outside: International Relations as Political Theory*. Cambridge: Cambridge University Press.

———. 1999. "Citizenship after the Modern Subject." In *Cosmopolitan Citizenship*, edited by Kimberly Hutchings and Roland Dannreuther. New York: St. Martin's Press.

———. 2010. *After the World, Before the Globe*. London: Routledge.

Yack, Bernard. 2012. *Nationalism and the Moral Psychology of Community*. Chicago: University of Chicago Press.

Young, Iris Marion. 1989. "Polity and Group Difference: A Critique of the Ideal of Universal Citizenship." *Ethics* 99 (2): 250–74.

———. 1997. "A Multicultural Continuum: A Critique of Will Kymlicka's Ethnic-Nation Dichotomy." *Constellations: An International Journal of Critical and Democratic Theory* 4 (1): 48.

———. 2011. *Justice and the Politics of Difference*. Princeton, NJ: Princeton University Press.

Young, Julian. 1997. *Heidegger, Philosophy, Nazism*. Cambridge: Cambridge University Press.

———. 2002. *Heidegger's Later Philosophy*. Cambridge: Cambridge University Press.

Yuval-Davis, Nira. 1997a. *Gender and Nation*. London: Sage.

———. 1997b. "Women, Citizenship and Difference." *Feminist Review* (57): 4–27.

Zerilli, Linda M. G. 2005. *Feminism and the Abyss of Freedom*. Chicago: University of Chicago Press.

Žižek, Slavoj. 1999. *The Ticklish Subject: The Absent Centre of Political Ontology*. New York: Verso.

Index

www.ingramcontent.com/pod-product-compliance
Ingram Content Group UK Ltd.
Pitfield, Milton Keynes, MK11 3LW, UK
UKHW012040150926
475560UK00001B/99

9 781783 488919